Building and Construction
Desk Book—with Forms

Building and Construction Desk Book—with Forms

Truman W. Cottom

Prentice-Hall, Inc.
Englewood Cliffs, New Jersey

Prentice-Hall International, Inc., *London*
Prentice-Hall of Australia, Pty., Ltd., *Sydney*
Prentice-Hall of Canada, Ltd., *Toronto*
Prentice-Hall of India Private Ltd., *New Delhi*
Prentice-Hall of Japan, Inc., *Tokyo*

© 1974 by
Prentice-Hall, Inc.
Englewood Cliffs, N.J.

All rights reserved. No part of this book may be reproduced in any form or by any means, without permission in writing from the publisher.

Library of Congress Cataloging in Publication Data

Cottom, Truman W
    Building and construction desk book--with forms.

    Bibliography: p.
    1. Construction industry--Management. I. Title.
TH438.C69   658'.92'4   73-12839
ISBN 0-13-086108-1

The reader is advised to consult the Contract Work Hours and Safety Act, as well as the Occupational Safety and Health Act, to determine if there is any regulation which may apply to the substances, devices, or procedures described herein.

This publication is designed to provide accurate and authoritative information in regard to the subject matter covered. It is sold with the understanding that the publisher is not engaged in rendering legal, accounting or other professional service. If legal advice or other expert assistance is required, the services of a competent professional person should be sought.

    . . . From the Declaration of Principles jointly adopted by a Committee of the American Bar Association and a Committee of Publishers and Associations.

<u>Printed in the United States of America</u>

Dedicated to the women in construction —among whom is my faithful and diligent wife, Zaidee, whose contribution to this book has been of major significance.

ABOUT THE AUTHOR

Truman W. Cottom is a consultant to contractors, architects, and engineers, and has provided information to the construction and building industries for many years. He has written and published hundreds of special reports dealing with recommended solutions to problems in these industries, and has worked with more than 8,000 contractors and sub-contractors. At present he is concentrating on assisting contractors in complying with the Occupational Safety and Health Act by making personal inspections and providing recommendations for compliance. He is also engaged in methods engineering studies to provide better and lower cost construction methods. Mr. Cottom is also the author of *Contractor's Desk Book* (Prentice-Hall, 1964).

Mr. Cottom provides information services for the building and construction industries, the most recent of which is the Architectural Research program handled by correspondence.

A Word from the Author

As a consultant who recognizes the innumerable problems that confront building and construction contractors, I have crystallized my experience and know-how in solving these problems and present this information for the first time in this volume. Over the years, these ideas and solutions have helped scores of companies save many hundreds of thousands of dollars.

This book will show you how you can set up definite and positive policies and procedures (Chapter 3) so that your projects will run smoothly without tying you down to frustrating details and constant decision making. It will show you how you can anticipate problems and how to apply quick accurate solutions (Chapters 7 and 11).

Many office and job-site forms—more than fifty in number—are included to save your time and provide the essential communications you need. These will be found in almost every chapter in the book.

In working with thousands of contractors nationwide for more than twenty years, I have had an opportunity to compare their operations and to quickly spot variations from normal. Invariably I discussed with the contractors the problem of shrinkages of inventories of materials and supplies. These losses were always alarming. In most cases, the contractor was losing 10 to 20 percent of his profit due to theft, improper handling and storage, and lack of control over materials and supplies. I have seen some instances where more than half of the profit disappeared into thin air by the time the project was complete! Chapters 8 and 9 give you the methods used to gain control over this extravagant waste on the job.

You can save plenty of headaches by applying the advice given in Chapter 10 concerning Occupational Health and Safety. The laws are digested and recommendations are made for compliance.

In addition, I have dealt with such critical areas in the building and construction industry as:

- Vandalism, carelessness, theft, and other unnecessary losses.
- Costly turnover of key management and supervisory personnel.
- The proper way to record transactions. When coupled with the proper forms, increased efficiency and lower operating costs almost always result.
- Because of its importance, I have dealt with computer estimating techniques and procedures which can reduce the elapsed time from as much as two weeks to a dramatic two days.

This brief word scarcely touches on the comprehensive matters that I have endeavored to cover for your benefit in this volume.

If you will apply these practical techniques and methods, I am confident you'll gain many of the same savings and increased efficiency that I've achieved during my years as a consultant.

Truman W. Cottom

TABLE OF CONTENTS

A Word from the Author **11**

1. Analyzing the Objectives of Your Organization **21**

2. Organizing and Managing Your Manpower to Produce an Efficient Organization **25**

How to Find Qualified Management People 26

   1. What Are the Qualifications of a Manager? • 26
   2. Does the Job Require Experience? • 27
   3. Does the Job Require Mechanical Skills? • 27
   4. How Much Responsibility Will the Job Carry? • 28
   5. Does the Job Require Creativity and Innovation? • 28
   6. How Much Stress and Pressure Does the Job Produce? • 28

Fitting New People into the Organization 29
How to Keep Good People 29
Management Development 30

   Setting Up a Procedure Manual • 30
   Making a Policy Chart • 30
   The Organization Chart • 30

Importance of Written Job Descriptions 31
Employment Application Forms 31

   Application for Office Positions • 31
   Telephone Check of Applicants • 41
   Selection and Evaluation • 41
   Determining Mechanical Aptitude • 41

How to Cope with Tensions 41

3. Internal Planning and Control of the Project for Lower Construction Costs ... 47

Planning Techniques 48

What Is the Best Planning Approach? • 50

Critical Path Method 50

Who Should Be Using CPM? • 51
What Size Contractors Can Best Utilize CPM? • 52
Types of Projects Applicable to CPM • 52

How to Obtain Cooperation from Subcontractors 53
What You Can Do to Avoid Delays Caused by Carriers 58
Forms and Procedures for Handling Bids 58

Abstract of Bids • 58
Record of Bidders • 64
Confirmation of Telephone Quotation • 64
Shop Drawing Approval • 64
Progress Reports • 65
Construction Cost Estimate Forms • 65
Joint Venture Agreements • 65

4. Utilizing the Computer to Reduce the Cost of Estimating, Scheduling and Designing 75

Estimating by Computer 75
Computer Application to Scheduling 76
Subcontractors and the Computer 76
Advantages of the Computer in Designing 77
Computer Drafting 77
Engineering Applications of the Computer 78
The Role of the Computer in Management 78

5. How Visual Control Systems Can Assist in Communicating Your Ideas 79

Visual Aids 79

Pressure-Sensitive Transfers • 79
Laminated Plastic Sheets • 80
Magnetic Visual Control Systems • 80

How Costs Can Be Lowered by Methods Engineering 80

Time Lapse Photos • 81
Video Tape Recorders and Monitors • 83

6. Streamlining Your Office Procedures and Forms to Reduce Errors and Save Time ... 85

Job Cost Distribution 85
The Intangible Benefits of Color Coding 86
You Can Design and Lay Out Your Own Business Forms 87
How to Recover a Letter After It Is Mailed 88
Communications Forms 89

7. Developing Tight Job-Site Project Control ... 91

Anticipating and Preventing Materials and Supplies Problems 91
Be Alert to Deviations Away from Schedule 92
What You Can Do to Keep the Project on Schedule 92
Provide Full Information to the Work Crews 93
Attitudes Are Contagious 93
How to Prevent Accumulation of Problems 95
Work Orders, Change Orders and Extra Work Orders 96
How to Stay Out of Hot Water 96

    Daily Construction Report Forms • 99
    Labor Report Forms • 99
    Employee Time and Job Forms • 105
    Pocket Size Data Systems • 106

8. Purchasing, Expediting, and Preventing Loss of Materials ... 117

Fire-Retardant Wood 117
Safeguards in the Purchase of Materials and Components 118
Storage and Protection of Materials 118
How to Reduce the Loss of Materials 119
Lumber and Millwork 120
Placement of Materials at the Job Site 122

    How to Avoid Delays Due to Materials and Supplies
        Problems • 123
    Materials Checklist • 123

Unauthorized Substitution of Materials 123
Proper Allowance for Loss When Figuring Materials 124
Protective Coatings for Steel 124

9. How to Get More from Your Tools and Equipment ... 127

Five-Step Procedure for a Well-Planned Equipment Program ... 127

Equipment Records • 128
Equipment Charges to Projects • 131

How Proper Training Can Reduce Equipment Costs and Improve Efficiency ... 131
Psychological Factors Affecting the Life of Equipment ... 133
Improving Efficiency of Equipment ... 134
Reducing Loss of Tools, Equipment and Materials ... 134
Reducing the Cost of Trucking Equipment ... 136
Operator and Craft Schools ... 143
Use of Helicopters in Construction ... 143
Application of the Laser Beam to Contracting ... 144
The Claw ... 144
Underground Piercing Equipment ... 144
How to Break Up Boulders and Concrete ... 145
Improving Your Woodworking Operations ... 145

10. Complying with the Occupational Health and Safety Requirements ... 147

Who Is Affected by the Act? ... 147
Complaints of Violations ... 148
Enforcement ... 148
Penalties for Violations ... 148
Records You Must Keep ... 149
Employers' Obligation to Employees ... 153
New Assistance from the Small Business Administration ... 153
OSHA Publications ... 153
Effects of Noise on Construction Workers ... 155
Clean Air Requirements ... 155
Fire Protection ... 156
Signs, Signals and Barricades ... 156
Materials Handling, Storage, Use, and Disposal ... 156
Disposal of Waste Materials ... 157
Hand and Power Tools ... 157
Electrical Requirements ... 157
Ladders and Scaffolding ... 158
Cranes, Derricks, Hoists, Elevators, and Conveyors ... 158
Motor Vehicles and Mechanized Equipment ... 158
Excavations, Trenching and Shoring ... 159
Concrete, Concrete Forms and Reinforcing ... 159

TABLE OF CONTENTS 17

    Steel Erection 159
    Tunnels and Shafts 160
    Compressed Air 160
    Demolition 161
    Blasting and the Use of Explosives 161
    Sources of Assistance on Safety and Health 161
    Effective Dates 162
    People Create Their Own Hazards 162
    The Hazards of Using Unsafe Equipment 163
    Some Underlying Causes of Accidents 163
    Psychological Factors Affecting Safety 164
    Accident Reports 164
    Keeping Up to Date on the Law 164

11. How Other Contractors Have Solved Tough Technical Problems 173

    How to Develop Your Problem-Solving Capability 173
    How to Lower Heavy Objects Without Handling Equipment 174
    How to Control Water Seepage Through Walls 175
    Controlling Cracks in Woodwork 176
    You Can *Prevent* Dusting of Concrete Floods! 176
    Methods of Curing Concrete 177

        Concrete Forming • 178

    Producing Correct Bushhammered Concrete Surfaces 178
    Removing Stains from Concrete 179
    Faster Cutting Through Concrete 181
    Solutions to Groundwater Problems 182

        Wellpoint Dewatering Method • 182
        Electro-Osmosis Dewatering Method • 183

    Flooring Problem Solutions 183
    How You Can Make Your Own Adobe Brick 183
    Salvaging Damaged Glass 184
    Improving Appearance of Acoustical Ceilings 184
    How to Keep Paint from Peeling 185
    Carpet Stains 185
    Preventing Birds from Roosting on Buildings 186
    Mortar Stains on Aluminum 186
    Tobacco Stain Removal 187
    How to Create Exposed Aggregate Concrete Surfaces 187

12. Sources of Information for the Construction Industries 189

    Associations 189
    Reference Sources 189
    Solutions to Problems in Building and Construction 190
    Building Code Organizations 190
    University Research Bureaus 190
    Problem Solution Associates 191
    Other Sources 191
    Federal Government 192
    Prentice-Hall, Inc. 193

        Personnel Management and Labor Relations • 193
        Tax Digest Service • 193
        Business Services • 193

    Stormproofing 194
    Computer Application to Construction 194
    Visual Aid Sources 194
    Methods Engineering Services and Equipment 195
    Critical Path and Other Scheduling Methods 195
    Sources of Supply for Charts, Graphs and Forms 195
    Concrete Forms and Form Liners 196
    Sources of Information on Concrete Problems 196
    Wellpointing and Electro-Osmosis 196
    Sources of Information on Fasteners 196
    Unusual Tools and Equipment 197
    Sources of Information on Management Development 197
    Information on Occupational Health and Safety 197

        Literature on Safety • 197

Appendix **199**

Index **217**

Building and Construction
Desk Book—with Forms

1

Analyzing the Objectives of Your Organization

Briefly stated, the basic objective of those engaged in management of a building or construction firm is to convert the ideas stipulated in a particular set of plans and specifications into reality. Although basic and primary responsibilities may rest upon the architect, engineer, or general contractor, responsibilities for accomplishment of most of the multitude of separate functions are divided among the various subcontractors, suppliers and service organizations.

The various subcontractors on a project, whether it is a house or a large construction project, have the same general objective: that is to accurately convert their work into functional reality. This simply stated objective is not so simple to accomplish within minimum time and with maximum efficiency. And since profits are determined by, and are in proportion to, the degree to which the individual contractor accomplishes the "minimum time—maximum efficiency" goal, it is extremely important that management people in the building and construction organization make a serious study of the steps necessary to attain reasonable proficiency in these areas.

That brings us to the primary objective of this book: *to assist those in the building and construction industries to accomplish their objectives in minimum time and with maximum efficiency.* And to accomplish our objective we will practice what we preach by presenting time-tested and proved methods, techniques and procedures, complete with appropriate forms so that you can get them in "minimum time and with maximum efficiency." Many thousands of you already have our *Contractor's Deskbook,** and you will probably agree that we pack a lot of valuable material into a small space.

The first step toward the simplification and application of the techniques and

---

* See page 191.

procedures necessary to attain your goal is to properly plan the fundamental direction you are to take; then break this down into separate related actions; then further reduce these into elementary individual actions. Obviously these must be placed in proper sequence.

The approach to sound development of the organization is spelled out in Chapter 2. The guidelines and proper forms are included. The step-by-step process of planning and scheduling each job or project is detailed in Chapter 3, Internal Planning and Control of the Project, and is continued in Chapter 7, Job-site Project Control, on to final completion of the project.

In any building or construction organization the degree of specialization of the management personnel will depend upon the size and scope of the firm's operations. This in turn will dictate the methods and procedures of operation. Although each firm is different, the fundamentals of organizing the job are essentially the same. Only the degree of detail breakdown varies with the size of the firm. It is this simple fact that makes the ideas contained in this book applicable to all firms engaged in any way in the building or construction industries.

Although many job functions may be performed by a single person, it is not at all feasible for one person to attempt to accomplish personally all of the hundreds of separate actions and functions required to bring even the simplest project to completion. Even the so-called "one-man organization" utilizes the services of others to handle accounting, legal, and usually bookkeeping and other specialized functions.

To determine the work load of any individual, we must analyze his daily activities. Forms, several examples of which are reproduced in Chapter 2, are available for this purpose. If the individual has time to sit over several cups of coffee and chat about things completely irrelevant to any business or job matter, he is probably not organized in line with our basic objective: to accomplish the job with maximum efficiency in minimum time. On the other hand, if he is so pressed with work that he suffers losses due to mistakes or inability to get to important matters, then he probably needs to organize his work load and possibly delegate some of it to others.

It it suggested that each individual within the organization, and especially those in top management, constantly keep this question before him: "Am I performing this job with maximum efficiency in minimum time?" If the answer is "No" or if there is any question about it, a determination should be made at once regarding the procedure or the steps necessary to attain this fundamental objective. Then, and only then, can the firm attain peak operating efficiency and peak profits.

Let's not take the attitude of the farmer who listened to the implement salesman's pitch and then remarked, "I'm not operating half as efficiently as I know how to now." Many people are a slave to habit. They admit that they know how to do better, but have no excuse except "that's the way we've been doing it." So let's ditch those obsolete habits when we find better ways. All of the ways suggested here are not going to be better than yours. Some of yours may be better than ours. So keep your best methods and capitalize on those of ours which you feel are better. Make sense?

The quality of organization in a firm can be quickly appraised by a look at the company's forms and records. It is not possible to have good, efficient organization without good, effective business forms. Therefore, proper forms for each purpose are a must.

After you have made a company procedure manual setting down the company's objectives, description of each job, and other pertinent data, the next step is to prepare the proper business forms. Many of these forms are already in print and can be obtained from the various sources listed throughout the book. These sources are also listed in Chapter 12, Sources of Information for the Construction Industries, or in the Appendix. Most of them publish catalogs showing specimen forms.

This is a good place to start. After using the forms you may find certain changes or additions desirable. Then you can have your printer make forms more nearly fitting your needs. You may wish to lay out your own forms on the drafting board as I have done for many of the forms shown in this book. This saves expensive typesetting. You can type the headings and simply paste them in on the form. On forms with considerable detail, it is usually better to make your layout one and a half to two times the desired size. The printer can reduce it to the desired size when he makes the offset plate.

For many years we have used an IBM Executive typewriter with upper and lower caps to make headings which look as good if not better than an expensive typeset job from the printer. The Follow-Up form shown in Figure 3-1 (page 49) is an example.

Now let's take a look at the forms needed in the office for your internal control. In summary these are:

1. Administrative (financial and management)
2. Personnel
3. Bookkeeping and clerical
4. Communication forms
5. Office procedure forms
6. Estimating forms and records
7. Contract and bid forms
8. Equipment records
9. Miscellaneous charts, graphs, etc.

This book contains forms in each of these important categories shown and discussed in the appropriate chapter.

Before we dive into job planning and job control, which is the subject of Chapter 3, we will want to examine the capabilities and limitations of the firm's key individuals. This brings us to the subject matter of Chapter 2.

2

Organizing and Managing Your Manpower to Produce an Efficient Organization

Even those contractors who sub out the maximum amount of their work must have some type of organization. Consequently, skill in organizing and managing human resources is vital to the success of every construction firm. Unfortunately, most of the literature on this subject assumes that the reader is in the manufacturing business, and comparatively little has been written with the contractor's problems in mind. This fact has caused many to lose interest in this important subject.

The problems of hiring skilled tradesmen are well known. For contractors employing union help, the agony of going through the whole waiting list from the union hall is something with which they have learned to live. Of course, the situation varies by locality, but it is unpleasant and costly at best. Many contractors have managed to hold on to the best men in spite of the many handicaps and obstacles. The question is how is this accomplished.

First of all, the old "bull-of-the-woods" technique went out with the horse and buggy. People no longer stand still for crude, undignified handling. Many of the old superintendents and foremen still have not improved upon those tactless methods of dealing with people. The obvious result is high labor turnover, higher than necessary absenteeism, and lower productivity. I have made a study of almost every conceivable type of problem arising in the building and construction business, and in more than 90 percent of these problems human relations factors were predominant! In looking behind the scenes in accident cases, I have found that the contributing causes were again largely human relations factors.

Few management people are fully aware of the high impact of these forces.

People are people regardless of where they work. They have remained the same throughout history. The advice of Plato, Aristotle, Plotinus, Thomas Aquinas and other ancient philosophers can be applied equally well to modern day human relations. Therefore, most of the problems can be circumvented by the application of sound, tested and proved *fundamentals of human motivation.*

As we know in the construction business, capable people are seldom out looking for a job. They are not down visiting the employment agencies. The question, then, is how are we going to locate talented and capable people? A study of some of the best organizations reveals that many of the most capable people were acquired by a constant awareness of the need and a constant search among all contacts for potential candidates.

Construction managers may be unable to hire the workmen they would like because of the union system. Instead of conducting a search for talented craftsmen in various fields, as my father was able to do back in the twenties, we generally have to hire craftsmen through the union hall. When the contractor does manage to build up a good force, his biggest problem becomes one of maintaining a fairly level volume of work to avoid having to lay people off at times. The firm which has a reputation of keeping its crews together will attract the best people. Some construction firms have actually gone into the manufacturing business in order to keep their best men busy during slack times. These manufacturing operations may include the making of trusses, pre-fab wall sections (or even pre-fab buildings), tilt-up wall sections, and numerous other operations of this type.

HOW TO FIND QUALIFIED MANAGEMENT PEOPLE

Obviously, we can't organize human resources until we have the people, so the first consideration is how to obtain the "right" people. Attracting and selecting people who will fit into your scheme of things—and be happy in their spot—is difficult and yet is of utmost importance. Keep your eyes open at all times for possible candidates for openings in your organization. Observe each candidate carefully in his present work and invite him in for a discussion. He may not be doing the type of work you want him to handle, but his basic qualities of intelligence, reliability, honesty, application of effort, and other important assets are more important than experience in many instances. Make a list of the important assets needed for the particular job and carry this list with you (possibly on a 3" x 5" card) so that you are constantly reminded to be on the watch for this type of individual.

Ask yourself the following questions to determine the qualities required for the job:

1. What Are the Qualifications of a Manager?

There are several qualities a manager will need. You will want to check into his ability to communicate and to motivate. Communications is discussed elsewhere, so let's talk about motivation. The term is usually applied to subordinates; however, the motivation of others, including superiors, is of equal importance. By this type of motivation *we do not mean manipulation*. Manipulation is *obtaining one's desires by misrepresentation or other devious means*. Most of us resent the manipulator, and quickly become aware of his tactics and develop a distrust of all his actions. Motivation is also separate from respect. A person may be able to gain respect but still not be able to motivate, or he may be able to motivate and not be able to gain respect.

A person's ability to motivate may be determined from the interview and from references. Use the following checklist.

- A. What is his previous record of accomplishment?
- B. How did he proceed in getting his job accomplished?
- C. How did he handle other people?
- D. What did he do to increase the efficiency of his group?
- E. Did he delegate too much or too little authority to others?
- F. Did he recognize individual differences and handle them accordingly?

2. Does the Job Require Experience?

If so, a thorough investigation into the past experiences of the applicant is important. Be sure to check more than one reference. The proper use of good forms and techniques will be most rewarding. Good forms can provide a checklist of the numerous points which must be investigated. Consider each point carefully, remembering that some people are quite capable of making themselves look good and manage to get along from one job to the next on their salesmanship. The downright misrepresentation made by some applicants is unbelievable! In such an instance, check with others who have had experience with the applicant, and ask each reference, "Who else knows him?"

3. Does the Job Require Mechanical Skills?

If so, give a simple mechanical ingenuity test. Don't take anyone's word for his mechanical abilities. We have heard an applicant spiel out glowing descriptions of his mechanical abilities only to find later that he had little or none. (He made a bookrack in school and received glowing praises from parents and friends!)

How quickly does he learn? This is a tricky one! Don't brand him as stupid too quickly. All of us appear stupid to experts in their field (sometimes even other experts)! An intelligence test is recommended here. A typical test of this type is the Wesman Personnel Classification Test.*

What is his capability in solving problems? This is a highly desirable quality for most jobs in the building and construction industries.

---

* Available from the Psychological Corporation, 304 East 45th Street, New York, N.Y. 10017.

4. How Much Responsibility Will the Job Carry?

Check into references; also ask the references for *names of other people* who know the individual. You'll be surprised what this technique will turn up! The "Application for Employment" forms included in this chapter include the proper questions to ask his references. If he is to be entrusted with keys and money, find out if he is bondable. Don't wait until he is hired to check on this. Is he or she emotionally capable of handling the responsibility of the job?

Responsibility places pressures on some people. You will need to know the individual's past performance in carrying responsibilities. Most jobs of responsibility require the ability to make decisions with good judgment. In both the interview and check of references, inquiry should be made into the applicant's capability to make decisions and his record of success.

In evaluating information from references and from the interview be sure to take into consideration the circumstances under which the applicant was working. Were his superiors unreasonable in their demands on him? Was he in the wrong job? The answers to these questions are especially important in cases where reports on his capability in decision making are unfavorable. He may have been working under conditions which would not yield a favorable report on anyone! A job with responsibility requires courage, and sometimes plenty of it! A person may be decisive and yet not possess courage. Is he submissive, or does he take a vigorous stand for what he thinks is right? Does he give up in the face of normal obstacles, or does he search for a way around these obstacles?

5. Does the Job Require Creativity and Innovation?

Creativity can be defined as the conception and generation of ideas. *Innovation* is the ability to put the ideas into practical application. Some individuals can turn out a steady stream of good ideas but are unable to put any of them to work. On the other hand, some people grab onto other people's ideas and quickly put them into profitable application. The true innovator does both. He generates good ideas and puts them to work. So, the *innovator* is usually the man you will be searching for. Don't forget this and make the mistake of falling for the idea man. His ideas will sound terrific but they may not be practical, and he may not be able to accomplish anything at all with them. This type of individual can certainly cost you plenty of money if you don't recognize him. On the other hand, if you can afford to employ an "idea man" as such, possibly you can make good use of him.

6. How Much Stress and Pressure Does the Job Produce?

This is a very important aspect to remember in evaluating a person for a job in the construction industry. Jobs in building and construction often produce more pressure than an applicant may have been subjected to previously. Make a thorough check on this capability when talking to references.

Obviously, it is not likely that any person who is being considered for a management position in a building or construction firm will fulfill all of these desirable qualifications. And if you find such a man he is probably a $30,000 to $100,000 per year executive!

The series of questions previously outlined is a guide to what is desired and a goal for all management people in the organization. Scores of books could be written expanding and detailing the fundamentals just covered, but I feel that sufficient material has been presented to enable you to accomplish your objectives.

FITTING NEW PEOPLE INTO THE ORGANIZATION

Obviously, there is a great difference between adding a craftsman to the work force and placing a new superintendent or other management employee on the payroll. But in any case, a new employee deserves some attention and assistance in getting started. Too many managers place a new person on the job without properly describing his duties or introducing him to the people with whom he must work. Seldom is a worker told whom to see about his various needs, and seldom is his job adequately described. Some construction foremen, superintendents, and other management men feel that it is sufficient to take the new man over to the person he is to work with and say, "You can work with Joe here. He'll get you started." Considerably more effort is needed to get a new person started. If he is not properly indoctrinated, he will be confused about too many things and his work will inevitably suffer. Also, remember that a new person is much more accident prone. He is almost always under a certain amount of pressure, whether his superiors are aware of it or not. This state of anxiety makes him more accident prone. Also, new environmental conditions are unfamiliar to him and present a greater hazard than they do to the older workers who have had the opportunity to gradually acclimate themselves to the conditions.

HOW TO KEEP GOOD PEOPLE

Recently I was visiting a general contractor who, quite naturally, had several items which required his attention. He jokingly remarked to his associate that he would have to "kick him out of the office." That remark inadvertently cost the contractor an important relationship of inestimable value. Joking can have its merits—if it does not degrade, insult, or humiliate others. Then it can be costly. This applies to all relationships with others. Don't forget that the most capable people are usually the most sensitive. One careless remark may not break up the relationship, but it is damaging. The result can be lower morale which in turn can infect others within and outside the organization. Remember these three basic needs: for security, for self-approval and for social approval. Make every effort to satisfy these needs and you will be able to keep your organization healthy and intact.

Written messages are even more likely to miss the mark unless well worded, spelled out, and well thought out. Consider the type of people who will receive your message and write it so that they will have no difficulty in comprehending its full meaning. If you have doubts about your writing ask someone else to read it and tell you what it says. You will often be surprised at his interpretation!

As we all know, misunderstandings can result in very costly errors, serious emotional disturbances, and countless loss of time, materials, and intangible costs. Many supervisory people feel that it is a waste of time to apply their efforts and their thinking toward the development of human relations. Actually, their job problems could be greatly lightened if more time and effort were devoted to organization and management of human resources.

MANAGEMENT DEVELOPMENT

Several good management development programs are available. These programs usually require only a few minutes of time periodically—weekly, biweekly or monthly. Investigate them. It may be one of the best investments in time and money you could make.

Setting Up a Procedure Manual

It is suggested that top management set up a company procedure manual in which all duties of individuals are decided upon and put in writing. The manual must then be followed closely, or a top level decision must be made to alter the job descriptions. As new jobs are created, it may be necessary to alter the job description so that everyone knows his own duties and his own authority. Don't forget to assign authority necessary to perform required duties! People certainly can't be expected to perform their assignments without the corresponding authority. One of the biggest mistakes made in organizational structures is that people are assigned definite responsibilities, and yet do not have the authority to take the actions necessary to carry out these same responsibilities.

Above all, do not have orders coming from several sources. Set up definite areas of responsibility and do not permit others to move in and issue direct orders to workers not under their direct supervision.

Making a Policy Chart

Your procedure manual should contain a policy chart outlining all company policies. Since every contingency cannot be anticipated at the start, make it as nearly complete as you can and add to it, or correct it, as different situations arise. Be sure to have everyone in management read and initial each policy and each change. Correct all individual deviations immediately and keep everyone on the right track. If people are allowed to drift off course gradually, the entire program will be lost.

The Organization Chart

Make the organization chart to fit your own operation. No standard organization chart will apply since each operation is different. It is good, however, to use a model chart as a general guide.

IMPORTANCE OF WRITTEN JOB DESCRIPTIONS

Figure 2-1 Shows a "Statement of Supervisory Expectancies" form ER-602 which can provide a guide for writing a job description for supervisory people. Everyone is entitled, first of all, to know exactly what is expected of him, what he is permitted to do, and the procedure he should use in accomplishing the many phases of his job. The only way to avoid confusion, frustration, anxiety and other deterioration in morale is to spell out clearly, at the start, the complete job description.

Do you presently have detailed written job descriptions on the various jobs in your organization? Most do not and all should have! These two-color forms are available for a few cents each. Dartnell's "Position Analysis" form JA-601, shown in Figure 2-2, may be used as a checklist of information essential to writing a good job description for any type of job, including executive, project manager, engineer, office manager or other.

These forms are only a few of the many forms available from the Dartnell Corporation.

EMPLOYMENT APPLICATION FORMS

On the following pages are shown several "Application for Employment" forms. The briefest of these is shown in Figure 2-3. It is a 5" x 8" card printed on both sides. Since only the bare essentials are included, a more detailed form is usually desired.

The "Application for Position" form OA-201 shown in Figure 2-4 is a general type of form which may be used for any position. For obvious reasons, specific forms for each type of job are highly desirable. The Dartnell Corporation has several forms designed for executive positions. Form EA-301 (not shown) is a four-page form containing what we consider the minimum desirable information to be obtained in the hiring of executives. To supplement the information obtained on this form, another Dartnell form, EP-302-R (not shown), should be used at the time of personal interview with the executive applicant. This form, "Patterned Interview Form—Executive Position," is a six page, two-color, detailed form. Very thorough, it provides a checklist of information considered essential in the appraisal of applicants for important positions.

The importance of good hiring and interview procedures cannot be overemphasized. One mistake in placing a person in an important position can cost thousands of dollars. I have seen numerous instances in which the company was wrecked by the careless hiring of the wrong person for an important position.

Since some of the executive application forms are six or eight pages long, they are not reproduced here.

Application for Office Positions

A two-page form, No. OA-205, shown in Figure 2-5, obtains the information needed in the hiring of people for office positions.

STATEMENT OF SUPERVISORY EXPECTANCIES

(Performance Review)

Date_____

Position_____Dept._____

Name_____Title_____

Supervisor_____
                                          Name and Title

Period covered: From_____To_____

Responsibilities relative to:

1. Personnel and supervisory relationships_____
      who will be supervised; what recommendation can be made with regard to promotions, releases, wage

_____
adjustments; training, development of subordinates; discipline; what is authority to know salaries; what is authority to use personnel files, etc.

2. Production_____
      what production schedules are to be met, what quality specifications, what handling of maintenance will be required

3. Costs and expenditures_____
      what extra personnel may be used; what authority is given for allocation of overtime, direct expenditures, petty cash,

withdrawal from stores, issue of equipment, personal expense accounts, etc.

4. Reports and correspondence_____
      what reports are expected with what accuracy, format, timing, and deadlines; what authority is given to finalize and

distribute certain reports; what reports are to be approved by what superior or other person

Form No. ER-602   Copyright, 1964, The Dartnell Corporation, Chicago, Ill. 60640. Printed in U. S. A.
                                                    Developed by The McMurry Company

FIGURE 2-1

Statement of Supervisory Expectancies

5. Superior/subordinate relationships_____
   what checking is to be done with supervisor daily, weekly, etc.; to what extent is the supervisor to be kept informed;

   how much authority may the subordinate take without checking

   _____

   _____

6. Interdepartmental relationships_____
   what individuals are to be contacted on company business; what authority is given for use of staff services; what

   interdepartmental contacts are to be made individually, which are to be channeled through superior

   _____

   _____

7. Special authorities_____
   what authority has been granted for use of company cars; admission of visitors; issuance of information about plant and processes

   _____

   _____

   _____

8. Technical skills, knowledge of company policy, procedures, and processes_____
   how much self-improvement is expected; what knowledge related to the

   job; what knowledge of personnel and public relations policies is expected

   _____

   _____

9. Innovations, new programs, improvements_____
   what are the areas in which supervisor expects improvements; what new programs are to be initiated;

   what innovations are indicated or necessary

   _____

10. Habits and off-the-job relationships_____
    what is expected with reference to financial affairs; family; drinking, gambling, etc.

    _____

    _____

11. Other_____

FIGURE 2-1 (cont.)

POSITION ANALYSIS

Title of position_____

General Division of the Organization: ☐ Executive   ☐ Finance and Accounting   ☐ Engineering   ☐ Sales   ☐ Transportation

☐ Purchasing   ☐ Personnel   ☐ Other_____

Organization nature of the position: ☐ Line   ☐ Staff   ☐ Combined

Will report to: Mr._____Title_____Age_____

Will be located at: Plant_____City_____State_____

The position is: ☐ A new one   ☐ An established one   ☐ Established but has new factors

For what functions in the organization will the person in this position be responsible?_____

_____

_____

_____

Describe any responsibilities associated with the position such as: Travel, community activities, association activities, public speaking, etc._____

_____

Rate Range for this position: $_____ to $_____   Expected hiring rate $_____

Describe nature of pay incentive_____

To what positions would next promotion normally lead?   1._____

2._____   3._____

Which of these should be considered in the appraisal?   ☐ 1   ☐ 2   ☐ 3

If accepted candidate does his job well, how soon could he reasonably expect promotion?_____

_____

After initial training, how closely will he be supervised?   ☐ Hourly   ☐ Daily   ☐ Weekly   ☐ Monthly   Describe_____

_____

How much of the person's work will be checked by others?_____

Which of these qualities should the person have to a high degree?   ☐ Flexibility; accepting changes   ☐ Resourcefulness   ☐ Creativity   ☐ Empathy

Explain_____

What kinds of problems, immediate or long range, are inherent in this position?_____

Form No. JA-601

Copyright, 1958, The Dartnell Corporation. Chicago 40, Ill., Printed in U. S. A.
Developed by The McMurry Company

FIGURE 2-2

Position Analysis

ORGANIZING AND MANAGING YOUR MANPOWER 35

SUPERVISORY RESPONSIBILITY

How many employees will report directly to this person?_____   Indirectly?_____

Titles of all supervisors who will report to this person and number of employees reporting to each:

1._____   7._____

2._____   8._____

3._____   9._____

4._____   10._____

5._____   11._____

6._____   12._____

WHAT AUTHORITY WILL HE HAVE:	TO RECOMMEND		TO DECIDE AND ACT	
1. Allocation of his budget	☐ Yes	☐ No	☐ Yes	☐ No
2. Organization of his own section (split or combine sections)	☐ Yes	☐ No	☐ Yes	☐ No
3. Increase or decrease number of employees under his supervision	☐ Yes	☐ No	☐ Yes	☐ No
4. Release or demote an employee	☐ Yes	☐ No	☐ Yes	☐ No
5. Revise standards of quality or quantity of product or service	☐ Yes	☐ No	☐ Yes	☐ No
6. Revise work flow	☐ Yes	☐ No	☐ Yes	☐ No
7. Revise operating policies	☐ Yes	☐ No	☐ Yes	☐ No
8. Establish his own itinerary	☐ Yes	☐ No	☐ Yes	☐ No
9. Establish itinerary of those reporting to him	☐ Yes	☐ No	☐ Yes	☐ No
10. Expenditures outside of his budget	☐ $50 ☐ $250 ☐ $1,000	☐ $100 ☐ $500 ☐ $10,000	☐ $50 ☐ $250 ☐ $1,000	☐ $100 ☐ $500 ☐ $10,000

11. _____

12. _____

13. _____

Analysis developed by_____Date_____

First approval_____Date_____ Approved_____Date_____

FIGURE 2-2 (cont.)

APPLICATION FOR EMPLOYMENT Date________

Name________ Soc. Sec. No.________ Are you a citizen of the U. S.? ☐ Yes, ☐ No

Address________ City and State________ Phone________

Job applying for________ Years of related experience________

Rate expected $____ per____ Shift preference____

Date of birth________
Check your State law as to discrimination because of age.

Height____ Weight____ No. of dependents____ Their ages____

CHECK THE ONES WHICH APPLY TO YOU
☐ Single, ☐ Married, ☐ Widowed, ☐ Divorced, ☐ Separated
☐ Rent home, ☐ Own home, ☐ Board, ☐ Live with relatives

Names of relatives employed here________

CIRCLE HIGHEST GRADE COMPLETED Grade School: 1 2 3 4 5 6 7 8; High School: 1 2 3 4; College: 1 2 3 4

Other training________ Special skills________

Ever employed here before? ☐ Yes, ☐ No; When?________ What job?________

Have you served in the U. S. Armed Forces? ☐ No, ☐ Yes; Branch________ From____ To____

What physical defects do you have?________

In case of emergency, notify________
Name Address Phone

YOUR WORK RECORD—COVER YOUR LAST FOUR JOBS

Last or present employer	Employer's address		Kind of business
Your job	Your supervisor's name		His title
When did you start?	When did you leave?	Your starting rate $ per	Your rate at leaving $ per
Why did you leave?			

(Over)

YOUR WORK RECORD—Continued

Next to last employer	Employer's address		Kind of business
Your job	Your supervisor's name		His title
When did you start?	When did you leave?	Your starting rate $ per	Your rate at leaving $ per
Why did you leave?			
Employer before that	Employer's address		Kind of business
Your job	Your supervisor's name		His title
When did you start?	When did you leave?	Your starting rate $ per	Your rate at leaving $ per
Why did you leave?			
Employer before that	Employer's address		Kind of business
Your job	Your supervisor's name		His title
When did you start?	When did you leave?	Your starting rate $ per	Your rate at leaving $ per
Why did you leave?			

Your signature________

DO NOT WRITE BELOW THIS LINE

Date________ Interviewer________

Form No. OC-200

FIGURE 2-3

Application for Employment
(front and back)

APPLICATION FOR POSITION

Date_____

Name (print)_____ Home Tel. No._____

Present address_____ How long have you lived there?_____
               No.   Street   City   State

Previous address_____ How long did you live there?_____
              No.   Street   City   State

Position applied for?_____ Earnings expected $_____

PERSONAL

Date of birth_____19____   ☐ Single,   ☐ Married,   ☐ Separated   No. children_____Their ages_____
Check your State law as to discrimination because of age.
Height____ft.____in. Weight____lbs.   ☐ Engaged,   ☐ Widowed,   ☐ Divorced   No. other dependents____Ages_____
Are you a U. S. citizen? ☐ Yes, ☐ No   Date of marriage_____   Soc. Sec. No._____

Do you: ☐ Own your home?   ☐ Rent?   ☐ Live with relatives?   ☐ Board?   ☐ Stay with friends?   Other_____

(If you rent) What monthly rent do you pay? $_____ Do you own your furniture? ☐ Yes, ☐ No

Is your wife employed? ☐ No, ☐ Yes, part time, ☐ Yes, full time; What kind of work?_____ Her earnings $_____per____

Do you carry life insurance? ☐ No, ☐ Yes; Amount $_____

What physical defects do you have?_____

In case of emergency, notify_____
                        Name   Address   Phone

EDUCATION

Type of School	Name and Address of School	Courses Majored in	Check Last Year Completed				Graduate? Give Degrees		Last Year Attended
Elementary			5	6	7	8	☐ Yes,	☐ No	19
High School			1	2	3	4	☐ Yes,	☐ No	19
College			1	2	3	4			19
College			1	2	3	4			19
Graduate School			1	2	3	4			19
Business or Trade School			1	2	3	4			19
Corresp. or Night School			1	2	3	4			19

Scholastic standing in H. S.?_____ In College?_____

EXTRACURRICULAR ACTIVITIES (athletics, clubs, etc.)
(Do not include military, racial, religious, or nationality groups)

In high school_____   In college_____

Offices held_____   Offices held_____

SERVICE IN U. S. ARMED FORCES

Have you served in the U. S. Armed Forces? ☐ Yes, ☐ No; (If yes) Date active duty started_____19____

Which Service?_____ What branch of that Service?_____ Starting Rank_____

Date of discharge_____19____ Rank at discharge_____

Form No. OA-201   Copyright, 1964, The Dartnell Corporation, Chicago, Ill. 60640. Printed in U. S. A.
Developed by The McMurry Company

FIGURE 2-4

Application for Position

WORK HISTORY

List below the names of all your employers, beginning with the most recent: a. Employer's Name b. Address and Telephone Number	Kind of Business	Time Employed				Nature of Work	Starting Salary	Salary at Leaving	Reasons for Leaving	Name of Immediate Superior
		From		To						
		Mo.	Yr.	Mo.	Yr.					
1. a. b.										Name Title
2. a. b.										Name Title
3. a. b.										Name Title
4. a. b.										Name Title
5. a. b.										Name Title
6. a. b.										Name Title
7. a. b.										Name Title
8. a. b.										Name Title

Indicate by number _____ any of the above employers whom you do not wish us to contact. Ever bonded? ☐ No, ☐ Yes; On what jobs? _____

References (Not former employers or relatives)

	Address	Phone Number
1.		
2.		
3.		

Are there any other experiences, skills, or qualifications which you feel would especially fit you for work with this Company? _____

If your application is considered favorably, on what date will you be available for work? _____ 19 ____ Signature _____

APPLICANT SHOULD NOT WRITE BELOW THIS LINE

1 2 3 4: Comments _____

_____

Interviewer: _____

FIGURE 2-4 (cont.)

ORGANIZING AND MANAGING YOUR MANPOWER 39

APPLICATION FOR OFFICE POSITION

Date_____

Name (print)_____ Home Tel. No._____ In whose name?_____

Your maiden name_____ Soc. Sec. No._____

Present address_____ How long have you lived there?_____
    No.   Street   City   State

Previous address_____ How long did you live there?_____
    No.   Street   City   State

Position applied for?_____ Earnings expected $_____

Date of birth _____ 19___ ☐ Single, ☐ Engaged, ☐ Married, ☐ Widowed, ☐ Separated, ☐ Divorced
Check your State law as to discrimination because of age.

Height ___ft. ___in. Weight _____ lbs. Date of marriage_____ Date of separation_____ Date of divorce_____

Are you a U. S. citizen? ☐ Yes, ☐ No   Number of children___ages___   Number of other dependents___ages___

Do you: ☐ Own your home? ☐ Rent? ☐ Live with relatives? ☐ Board? ☐ Stay with friends? Other_____

Is your wife (husband) employed? ☐ No, ☐ Yes, part time, ☐ Yes, full time; What kind of work?_____ Earnings $_____ per_____

In case of emergency, notify_____
    Name   Address   Phone

EDUCATION

Type of School	Name and Address of School	Courses Majored In	Check Last Year Completed				Graduate? Give Degrees		Last Year Attended
Elementary			5	6	7	8	☐ Yes,	☐ No	19
High School			1	2	3	4	☐ Yes,	☐ No	19
College			1	2	3	4			19
Business School	A.								19
	B.								19
Corresp. or Night School									19

(Indicate below specific experience which you have had)

Check Here	Type of Experience	Yrs.	Mos.	Check Here	Type of Experience	Yrs.	Mos.	Check Here	Type of Experience	Yrs.	Mos.
	Addressograph Operator				Confidential Secretary				Office Boy		
	Blue Print Mach. Operator				Dictating Mach. Transcript'n				Office Supervisor		
	Clerical Supervisor				Key Punch Operator				Photostat Operator		
	Clerk				Mail Clerk				Receptionist		
	Correspondence				Duplicating Mach. Operator				Secretary		
	Cost				Ditto				Telephone Swbd. Operator		
	File				Mimeograph				Teletype Operator		
	General				Multigraph				Timekeeper		
	Statistical				Multilith						
	Stock				Other						

(Indicate below your office skills and check office machines you can operate efficiently)

☐ Typewriter   Speed in typing_____   ☐ Billing Machine   Which ones_____
☐ Electric Typewriter   Speed in typing_____   ☐ Bookkeeping Machines   Which ones_____
☐ Vari-type   Speed in typing_____   ☐ Accounting Machine   Which ones_____
☐ Shorthand   Speed in taking dictation_____   ☐ Calculating Machine   Which ones_____
☐ Stenotype   Speed in taking dictation_____   ☐ Tabulating Machine   Which ones_____

What other languages do you speak?_____ Read?_____

Form No. OA-205   Copyright, 1964, The Dartnell Corporation, Chicago, Ill. 60640. Printed in U. S. A. Developed by The McMurry Company

FIGURE 2-5

Application for Office Position

WORK HISTORY
(Record U. S. Military Service as a position)

List below the names of all your former employers, beginning with the most recent:
a. Employer's Name
b. Address and telephone number

Kind of Business	Time Employed				Nature of Work	Starting Salary	Salary at Leaving	Reasons for Leaving	Name of Immediate Superior
	From		To						
	Mo.	Yr.	Mo.	Yr.					

1. a. _____ b. _____ — Name / Title
2. a. _____ b. _____ — Name / Title
3. a. _____ b. _____ — Name / Title
4. a. _____ b. _____ — Name / Title
5. a. _____ b. _____ — Name / Title
6. a. _____ b. _____ — Name / Title
7. a. _____ b. _____ — Name / Title
8. a. _____ b. _____ — Name / Title

Indicate by number _____ any of the above employers whom you do not wish us to contact. Ever bonded? ☐ No, ☐ Yes; On what jobs? _____

References (Not former employers or relatives)

	Address	Phone Number
1.		
2.		

✓ If you now have children or housekeeping duties, how will these be cared for? _____

✓ What transportation would you use from home to office? _____

If your application is considered favorably, on what date will you be available for work? _____ 19____ Signature _____

APPLICANT SHOULD NOT WRITE BELOW THIS LINE

1 2 3 4: Comments _____

Interviewer _____

FIGURE 2-5 (cont.)

ORGANIZING AND MANAGING YOUR MANPOWER

Telephone Check of Applicants

It is essential to make a telephone check, long distance if necessary, to verify information given by applicants and to obtain additional information. A checklist for this purpose is essential. Form OT-203, shown in Figure 2-6, may be used in connection with office or average employee jobs; however, for executive applicants, use form No. ET-303, reproduced in Figure 2-7. Use of this form will bring out information absolutely essential in filling important positions.

Selection and Evaluation

It is important in the evaluation of an applicant for any position to differentiate between what the applicant *can do* and what he *will do* (or what he *probably* will do). The "Selection and Evaluation Summary" form No. ES-404R-3 (Figure 2-8) is helpful in making this appraisal.

Determining Mechanical Aptitude

Many jobs in the building and construction industries require some degree of mechanical aptitude and manual dexterity. Several mechanical aptitude tests are available from the Psychological Corporation (304 East 45th Street, New York, N.Y. 10017). The "Bennett Mechanical Comprehension Test" is one which measures the ability to understand mechanical relationships and physical laws in practical situations. Problems consist of drawings and simply phrased questions about them. A specimen set of two tests is available from the Psychological Corporation for one dollar.

Although there are several dexterity tests available, the applicant's capability can be easily determined by having him assemble some small project such as a mechanical toy, piece of machinery or specially prepared device. Several of these are shown in a test catalog available from the Psychological Corporation. The "Engineering and Physical Science Aptitude Test" described in the catalog is designed to measure mechanical and technical aptitude. It consists of six sections: Mathematics, Formulation, Physical Science Comprehension, Arithmetic Reasoning, Verbal Comprehension, and Mechanical Comprehension. A specimen set 3V807 is available from the Psychological Corporation for $1.20. Their catalog lists numerous other tests of various types. The firm has been in business since 1921 and is considered foremost in its field. Its tests are in general use in schools and industry.

HOW TO COPE WITH TENSIONS

One of the most important factors in the development and management of an organization is the ability to cope with the tensions that are inevitable in the building and construction business. Your success depends considerably upon your understanding of tension, how tensions are produced, and what remedial measures can be taken to reduce the pressures which cause *excessive* tension. I emphasize *excessive* because a

TELEPHONE CHECK

Name of Applicant_____

Former Supervisor_____ Title_____

Company Where Applicant Worked_____ Telephone Number_____

1. Mr. (name) has applied for employment with us. I would like to *verify* some of the information given us. When did he work for your company?   From_____ 19____ To_____ 19____

2. What was his job when he started to work for you? _____

   When he left? _____

3. He says his earnings were $_____ per_____ Is that correct?   ☐ Yes,   ☐ No, $_____

4. What did you think of him? (Quality and quantity of work, attendance, how he got along with others, etc.) _____

4a. What accidents has he had? _____

5. Why did he leave your company? _____

6. Would you re-employ him?   ☐ Yes,   ☐ No,   (If not, why not?)_____

Additional comments_____

Date of Check_____ 19____ Made by_____

Form No. OT-208

Copyright, 1959, The Dartnell Corporation, Chicago 40, Ill., Printed in U. S. A.
Developed by The McMurry Company

FIGURE 2-6

Telephone Check

ORGANIZING AND MANAGING YOUR MANPOWER 43

TELEPHONE CHECK ON EXECUTIVE APPLICANT_____
Name of Applicant

_____ _____
Person Contacted   Position

_____ _____
Company   City and State   Telephone Number

1. I wish to *verify* some of the information given to us by Mr. (name) whom we are considering for an executive position. Do you remember him? What were the dates of his employment with your Company? From_____19____To_____19____
   _{Do dates check?}

2. What was he doing when he started? _____
   _{Did he exaggerate?}
   When he left? _____
   _{Did he progress?}

3. He says he was earning $_____ per_____ when he left. Is that right? ☐ Yes, ☐ No; $_____
   _{Did he falsify?}

4. What was basis of his compensation? _____
   _{Any profit sharing? Bonus? Evidence of ownership?}

5. What did you think of him? _____
   _{Did he get along with his superiors?}

6. Did he have any supervision of others? ☐ No, ☐ Yes; How many?_____
   _{Does this check?}
   (If yes) How well did he handle it? _____
   _{Is he a leader or a driver?}

7. How closely was it necessary to supervise him? _____
   _{Was he hard to manage? Did he need help constantly?}

8. How willing was he to accept responsibility? _____
   _{Did he seek responsibility? Was he afraid of it?}

9. Did he have any responsibility for policy formulation? ☐ No, ☐ Yes; How much?_____
   (If yes) How well did he handle it? _____
   _{Good judgment? Realistic? Able to plan ahead?}

10. Did he develop or initiate any new plans or programs? _____
    _{Initiative? Creative? Realistic?}

11. How well did he "sell" his ideas? _____
    _{Self-reliance? Ability to adjust to others' needs?}

12. How hard did he work? Did he finish what he started? _____
    _{Is he habitually industrious? Persevering?}

13. How well did he plan his work? _____
    _{Efficient? Able to plan?}

14. How well did he get along with other people? _____
    _{Is he a troublemaker?}

15. How much time did he lose from work? _____
    _{Conscientious? Health problems?}

16. Why did he leave? _____
    _{Good reasons? Do they check?}

17. Would you re-employ him? ☐ Yes, ☐ No; Why not?_____
    _{Does this affect his suitability with us?}

18. Did he have any domestic or financial difficulties which interfered with work? ☐ No, ☐ Yes; What?_____
    _{Immaturity?}

19. How about drinking or gambling? ☐ No, ☐ Yes; What?_____
    _{Immaturity?}

20. What are his outstanding strong points? _____

21. What are his weak points? _____

22. For what type of position do you feel he is best qualified? _____

Checked by_____Date_____
Form No. ET-808   Copyright, 1959, The Dartnell Corporation, Chicago 40, Ill., Printed in U. S. A.
                                          Developed by The McMurry Company

FIGURE 2-7

Telephone Check on Executive Applicant

SELECTION AND EVALUATION SUMMARY

Applicant's Name_____Date_____19____
Position Applied for_____Job Class_____

CAN-DO FACTORS

Check Each Factor	Above Requirements	Meets Requirements	Marginal	Unacceptable
Appearance, manner....................................			
Availability..			
Education..			
Intelligence (as measured by test).....................			
Experience in this field (if applicable)...............			
Knowledge of the product (if applicable)...............			
Physical condition, health.............................			

WILL-DO FACTORS

CHARACTER TRAITS (Basic Habits)

	A Lot	Some	Not Much	Almost None
STABILITY; maintaining same jobs and interests............				
INDUSTRY; willingness to work..........................				
PERSEVERANCE; finishing what he starts.................				
ABILITY TO GET ALONG WITH OTHERS.................				
LOYALTY; identifying with employer.....................				
SELF-RELIANCE; standing on own feet, making own decisions				
LEADERSHIP..				

JOB MOTIVATIONS (not already satisfied off the job)

NEED FOR INCOME or desire for money.................				
NEED FOR SECURITY...................................				
NEED FOR STATUS.....................................				
NEED FOR POWER......................................				
NEED TO INVESTIGATE.................................				
NEED TO EXCEL (to compete)..........................				
NEED FOR PERFECTION................................				
NEED TO SERVE.......................................				

BASIC ENERGY LEVEL (vigor, initiative, drive, enthusiasm)

DEGREE OF EMOTIONAL MATURITY

Dependence..				
Regard for consequences...............................				
Capacity for self-discipline..........................				
Selfishness...				
Show-off tendencies...................................				
Pleasure-mindedness...................................				
Destructive tendencies................................				
Wishful thinking......................................				
Willingness to accept responsibility..................				

Important: Do not add or average these factors in making the Over-all Rating. Match the qualifications of the applicant against the requirements of the *particular position* for which he is being considered, and consider the importance of each mismatch.

Strong Points for This Position_____

Weak Points for This Position_____

Over-all Rating: [1] [2] [3] [4]   Recommendation to Employ: [ ] Yes [ ] No   Rating by_____

Form No. FS-404R-3   Copyright, 1964, The Dartnell Corporation, Chicago, Ill. 60640. Printed in U. S. A.
Developed by The McMurry Company

FIGURE 2-8

Selection and Evaluation Summary

ORGANIZING AND MANAGING YOUR MANPOWER

normal amount of tension is not only healthy but actually enables us all to produce more. Since this subject is so vital let's take a closer look at it.

First, let's see how these pressures build tensions. An average day for Mr. C. may go something like this: He gets up early so that he can meet Mr. X at 7:30 and get him started on the job which should have been finished last week. He gets into his car, only to find that the park lights were left on and the engine won't start. He then has to get his wife's car out of the garage and discovers that there may not be enough gas to reach the first filling station. Tensions already, and it is only 7:00!

Mr. C. arrives at the office after several frustrating delays in traffic only to find that Mr. X has not yet arrived. Did Mr. X forget about the early appointment? Has he been there and gone? Maybe he forgot about the entire job? More tension! He hears the phone ringing as he fumbles to get the office door unlocked. It is the bookkeeper. She is ill and can't make it in today. Mr. C. had already thought up several things he needed the bookkeeper to do right away. Now he will have to dig all around the office to try to get these details done. More tension! And it isn't even 8 A.M. yet.

So this is the way the day starts, and it may go on all day. By the end of the day Mr. C. is a wreck. If this pressure is allowed to build day after day, it is going to get him in the end—and the end may be sooner than anticipated.

We have all felt these tensions and pressures. The question is, what can be done about them? Fortunately, there *are* things that can be done.

It is recognized, first of all, that the very nature of the construction business makes organization difficult. New challenges constantly arise. Don't forget that it is these new challenges that make the business interesting (although we all wish sometimes that it wasn't quite so challenging). But there are preparations we can make to help ease the blow of these challenges. Many decisions must be made daily, and frequently there is one after the other throughout the entire day. Making decisions requires effort. That is why many people procrastinate in making decisions—it is much easier to put off the decision than to make it. You will also notice that few procrastinators have any degree of success (excepting, of course, a few politicians).

We can't avoid these decisions. They will have to be made, so the sooner the better. Otherwise, the accumulation of numerous problems hanging over, awaiting a decision, builds pressure and resulting tension which can only be relieved by getting them cleared. It is always a relief to make the decision, even if there is doubt about the correctness of that decision.

Another factor which is difficult to control is that most of our plans depend upon other people. The best laid plans go haywire when other people are involved. Most people do not plan well, if at all. Therefore they miss appointments and foul up their own plans, which in turn fouls up yours. Much of this can be avoided by good communications. When you are depending upon others, be sure to question them about how well *their* plans are made.

It is usually a good idea to look into the schedules of these other people and ask questions, such as: "How will you get there, Joe?" (This may reveal that his car is in the garage and he "hopes" it will be ready.) "Do you have the parts or material you need?" (This may reveal that he "hopes" it has been delivered to the job.) These questions are typical. Each situation requires investigation and planning. If you leave everything to chance, chances are things will not go smoothly.

Proper planning and organization can be the best way for you to reduce frustrations which cause tension. Always carry note paper and pen or pencil. Always have note paper available where you work, in all of your vehicles, and at home. When something comes to mind that requires action or investigation, make a note of it. As these actions are completed, scratch them off the list and consolidate the remaining items into a new list.

Complete as many details as possible as soon as possible. Keep them from piling up. A big backlog of things requiring action builds pressure. Learn to delegate, but also make a note to follow up to see that the job has been completed properly and on time. Devise some type of follow-up system such as that presented in the form shown in Figure 3-1 (page 49).

Avoid noisy, confused conditions where possible. Invariably, noise and confusion cause pressures. When conditions have a noticeable effect on nerves, do something about it before the buildup is severe.

Above all, don't take the easy way of handling the demands on your time by accepting obligations you shouldn't. Stop and think before you accept a questionable responsibility. Should someone else take care of it? Does it need to be done at all?

Up to now you may feel that we are giving you a personal self-improvement course. Perhaps so. But all of these principles apply to the entire management organization. After giving full consideration to them, the best way to convey these simple principles is by example. Then you can pass the book on to others—or you can extract them and use them as memos to others in the organization.

One last note on this subject. Prepare yourself with facts. If you are not sure, get the facts from reliable sources. This is essential in building confidence in yourself and gaining the confidence of others. Doing this can relieve frustrating pressures and avoid errors.

3

Internal Planning and Control of the Project for Lower Construction Costs

Success in the building and construction industries depends considerably upon how well management is able to plan and coordinate the numerous steps involved in the average project. Since each project is different, each one poses new and different problems. It naturally follows that success in planning the project depends upon how well management is able to *anticipate problems* and the *ingenuity* he uses in *solving* these problems. The construction business is largely a problem-solving business. One problem can be very costly, as we all know. Therefore, internal planning must be given serious attention and every effort made to anticipate problems before construction of the project begins.

Job superintendents are frequently very casual about their responsibility to manage the job efficiently. Consequently it is the imperative duty of top management, regardless of the size of the firm, to see that the person in charge of the project is thoroughly acquainted with the details of the project. Question him. Bring up some of the details for discussion. Let him know you aren't asleep on the project. You must set the proper example by efficient planning and follow-up. Only then will you be able to impress others with the necessity for proper scheduling, planning and follow-through.

Quite often, we look but do not see. We can not give proper consideration to the study of plans and specifications unless we are free from distractions and interruptions. Failure to concentrate at this important stage of the project can be very costly later. Many otherwise competent management people notice impending errors or omissions but fail to record them for proper remedial action. How many times have we heard, "I noticed that but forgot about it"? So the rule is: Make notes "on the spot" before you forget it. Try carrying a few 3″ x 5″ cards in your shirt pocket for making notes. Then, record these notes for follow-up until the job is accomplished.

An example of a follow-up form is shown in Figure 3-1. After setting up a system, a determined effort must be made to use it as planned until it becomes habit.

The entire management organization should be at least casually familiar with the company's scheduling system, even those whose primary function is finance, personnel, purchasing, expediting or other. The schedule, or CPM Master Chart, should be the *common denominator* of all management personnel in a construction organization. It graphically presents the primary goal of the company at the time. After all, an organization is a team, and a team must be coordinated. To be coordinated it is essential that every member of the team know the plays of the other members. That's why the football team goes into the huddle—to learn what play is coming up. And just like the football team, the management team must know all of the plays thoroughly. Otherwise team members run in all directions and utter chaos results. Haven't you seen this condition at times in your own organization?

You've probably seen that famous cartoon (showing two men leaning back in their chairs, their feet on the table and a drink in their hands), captioned: "Next week we're going to have to get organized." In the first place, you don't get organized suddenly. And you don't have to stop all operations and reshuffle your entire organization. We have all heard of management consultants who come in and tear the place apart, sometimes draining the firm of all of its operating capital. Obviously, we want to avoid this situation.

Organization is a state of mind. The top people must be constantly aware of the importance of good planning and practical procedures which avoid confusion, mistakes and inefficiency. The people within the company *can* and *should* organize their own operations. This is not to say that there is no place for consultants or specialists. Far from it. It is certainly advisable to make use of specialists when the firm is on unfamiliar ground. But if top management feels that some outside group of high pressure, generalized "management consultants" is necessary, perhaps the situation is already hopeless. For every firm which has profited from the "management consultant" take-over, *there are a thousand which have not!*

All that is needed are simple guide lines and the earnest desire and determination to solve one's own problems before they become too serious. A firm becomes hopelessly mired down if problems are allowed to accumulate. As I have said previously, organization starts at the very top. I have seldom seen an inefficient man at the head of a good organization. Where such a situation exists, the organization begins to deteriorate.

PLANNING TECHNIQUES

Considerable thinking and planning must be done before any job can be estimated and bid; otherwise, the risk is too great even to bid the job. Construction men have had to gain a mental picture of the entire project and keep this mass of data in mind somewhat as a computer does. There is no substitute for the human brain, but scientific methods of analyzing this information have now enabled construction men to

INTERNAL PLANNING AND CONTROL OF THE PROJECT 49

JOB NO	PROJECT	REASON FOR DELAY	PERSON RESPONSIBLE	DATE	ORDERED FROM	ITEM RECD.	WORK COMP.	REMARKS

FOLLOW-UP SUMMARY

FIGURE 3-1
Follow-Up Summary

get a better grasp of complex operations and to portray this information in logical form. Such a method is the Critical Path Method. It is a tool which simplifies the sequence of operations and requires us to think logically through the entire project.

What Is the Best Planning Approach?

One of the best methods is the *conference technique* in which key people concerned with the planning and execution of the project meet and discuss the entire procedure from start to finish. A large blackboard * is useful for making a rough diagram of the project as it evolves in the discussion.

If the Critical Path Method (CPM) is to be used, it is advisable to bring in a specialized consultant at least on the first project, to guide the discussion and make the arrow diagram needed. Much depends upon the complexity of the project. On a thirty-million-dollar hospital project, for example, the arrow network may contain 1,200 to 1,800 arrows. The time required to think through the entire project may be four or five days, but this time will be repaid, perhaps many times, before the project is completed.

CPM requires the detailed and orderly organization of all the steps in the building process. That is its main value. Many contractors are not using CPM because it takes longer to prepare. Since the use of CPM almost invariably is more than worth the initial investment of time, this objection is purely psychological. It is only natural to want to get the job under way as quickly as possible, and it is human nature to procrastinate.

We will not discuss all of the scientific scheduling methods here since most construction management people are familiar with at least one or more of them. Many have adopted the Critical Path Method of project scheduling and many more would profit by the application of advanced planning and scheduling methods.

CRITICAL PATH METHOD

One contractor on a multi-million dollar hospital was very pleased with the fact that he completed the project on schedule—and without the use of CPM or computers. What does this tell us? Nothing, except that ample time was allowed for the completion of the job. There is no way to compare it with another method of scheduling because no other project will ever be exactly similar. Even if we tore the building down and started all over on a new structure using CPM and computers the conditions would not be the same. My point is that completing the projects on schedule means little. All depends upon how lenient we are in setting schedules. Experience gained after several years of application of advanced scheduling techniques indicates that considerable time is saved on the average project. Some contractors feel that savings in time average as high as 20 percent. But even if project time is reduced by 5 percent, contractors will either have to get with it or eventually risk extinction. The handwriting is on the wall!

CPM is readily adaptable to construction projects of all types and sizes. Although it is not as complicated as it may at first appear, it is suggested that the firm send

---

* Also see Chapter 5, Visual Control Systems.

its key project planners to one of the several schools for a few days. This will ultimately be less costly than trying to overcome the many baffling questions which will continually arise. Another approach is to obtain the services of a good CPM consultant to work with key people in setting up a project on CPM. Then, if more education and experience are desired, one or more of the firm's key people may attend a CPM seminar.

Several good books are available on CPM and the various scheduling techniques. Some of these are reviewed in Chapter 12, Sources of Information for the Construction Industries. More information will be found in Chapter 7, Utilizing the Computer, and elsewhere in the book.

CPM provides a method of predicting which elements of a project are critical with respect to *time*. Time is our number one concern in planning and scheduling. Our objective is to get the job through to final acceptance in a minimum time at a minimum cost. CPM makes use of a network diagram in which each job or phase of operations is represented by an arrow. Time, usually in numbers of days, is shown on the arrow. This enables us to *determine in advance* which factors or phases of the job can hold up the project. If we recognize these and keep our attention focused on them, we can usually keep them moving at maximum speed. Obviously, there is no substitute for experience; without it we would seldom be able to anticipate factors which can develop too late to avoid delays.

During every construction project there are jobs which cannot be started until some job has been completed. We cannot start everything at once. Even in the highest of production operations this is true. For example, assembly operations can never be started until parts are made. So, on the construction project we must focus our attention on the items which must be completed before another item can start.

The "critical path," then, for any project, is the connected sequence of events which will require the *longest period of time*. This string of operations determines the *minimum completion time* of the project. *To reduce the completion time, one or more of these items must be accelerated.* However, if one of these items is shortened by working overtime or in some other way, another item, not on the critical path, may become *critical*.

Let me clear up one point right now. CPM can be used to advantage *without the application of a computer*. The computer is used mainly to *reduce the time in making lengthy computations*. On large projects it is an important tool.

The big advantage of the CPM chart is that a few minutes spent in studying the chart will reveal any dangers of potential delay. Changing situations always require a new appraisal of the consequences, but this is true of any type of scheduling method. Also, we know where to concentrate our efforts in follow-through to be sure all jobs along the critical path keep moving on schedule—or ahead, if possible. If we gain an extra day anywhere along the critical path, we have reduced the total project time by one day.

Who Should Be Using CPM?

Since the advent of CPM and other advanced scheduling methods, almost every construction man has given thought to the application of these techniques to his own operations. Numerous questions need to be answered. What types of projects lend themselves to CPM? On what dollar volume project can it be utilized to advantage? Will it be *necessary* to use CPM to be competitive? Is CPM really better? Will it actually result in dollars saved on the job?

In general, CPM can definitely be applied to advantage on large, complex projects, and many contractors feel that it has been an advantage on very small jobs, such as the construction of a single-family residence. Efficient application of CPM can only be gained after some education and experience with it. Obviously, some time and expense is entailed in learning the method, but in most cases contractors have found scheduling by CPM to be a great advantage over previous methods. Certainly it is more accurate and complete. Many contractors who are experienced with CPM now apply it to all jobs regardless of size. If the project is small, the schedule and diagram can be done quickly and without the application of a computer.

CPM is a *decision-making process*. It forces decisions on the front end where they *should* be made, instead of in the middle of the project where delays and indecision can be very expensive.

What Size Contractors Can Best Utilize CPM?

Although larger size projects are best scheduled by CPM, the smaller contractor who has an opportunity to follow the job closely can apply CPM more easily than the large organization. This is primarily due to the problem of communication. Some of the older superintendents seem to have closed their minds to newer scheduling techniques and many contractors have simply given up their efforts to persuade them to use such techniques. On the other hand, if the project manager is involved all the way through in the planning and follows through in the field, there is little problem. A contractor building jobs in the $100,000-range and up can apply CPM effectively. After considerable experience with CPM, many contractors apply it to almost everything and wonder how they got along without it.

CPM, as well as any other intense form of planning, costs more on the front end but pays back with dividends through shorter project time, fewer errors, and in other ways previously cited. CPM is certainly no panacea for all planning problems, but it is a scientific method which has now proved its value beyond doubt.

Types of Projects Applicable to CPM

On multimillion-dollar building and construction projects, CPM can certainly be used to advantage. First, consider the complexity of the project. If the job is simply an excavation operation or pipeline job, the advantages of CPM would be doubtful. On a complex mechanical or engineering project, CPM would have advantages. CPM has also been used by contractors in remodeling where numerous small jobs are constantly in progress.

INTERNAL PLANNING AND CONTROL OF THE PROJECT 53

HOW TO OBTAIN COOPERATION FROM SUBCONTRACTORS

Regardless of how well the superintendent handles his own forces, he must be certain that vendors and subcontractors maintain their own schedules. Recommendations for selecting subcontractors and suppliers is discussed elsewhere in this chapter. Also, suggestions have been made for impressing upon the subs and suppliers the seriousness of the planning and their importance in the general scheme of the operation. Still, we cannot relax and expect everyone concerned to perform like clockwork. This rarely happens. Therefore, the proper follow-through is needed. The person responsible for expediting must anticipate well ahead of time any possible reasons for delay and go to work early enough to avoid delays. Don't take anything for granted. Assume *nothing*. Get positive answers—even *proof* of the status of questionable schedules. A constant awareness and follow-up will aid in solidifying your determination to adhere to the schedule.

Perhaps one of the most significant developments in the evolution of the building industry has been the continuous progression toward specialization. Skilled and educated craftsmen are required to properly install and service the many diverse and complex components of today's structures, necessitating highly specialized building trade groups. The industrial revolution in other industries developed and made apparent the advantages of specialization. The successful completion of even the smallest project today depends upon the coordination between the various subcontractors and the general contractor.

Naturally the first basic element in a successful relationship between contractor and subcontractor is the capability of both parties. It is certainly a simple matter for each party to check into the business reputation of the other, and this should be thoroughly explored before entering into a business relationship.

The advantages of subcontracting are evident. Since the subcontract is taken at a fixed figure, the builder knows his costs and the subcontractor is encouraged to use every efficient means at his disposal to expedite the job. The supervision of labor, then, is left to the subcontractor, thus relieving the general contractor of this detailed responsibility. This simplifies the management problem for both the general contractor and the sub.

Now these same factors which encourage higher productivity by the subcontractors also lead in many instances to a reduction in the quality of the work. This is one area of considerable discord between contractors and subcontractors. The subcontractor makes a serious mistake if he plans to skimp on the quality of work or materials, and then later attempt to cover it up, in order to get the job. He should figure the job honestly and go over the details with the contractor, pointing out any possibilities for reducing cost. There should be an adequate written contract providing a thorough understanding between the two parties to prevent headaches and sources of misunderstanding. So, the first prerequisite to a smooth relationship is to put everything, regardless of how small, into writing. Not only the original contract but also any change orders or extra work orders should be in writing.

A sample Subcontract Agreement (form 5160) is shown in Figure 3-2. Quoting

SUBCONTRACTOR AGREEMENT
FORM 5160

THIS AGREEMENT made and entered into this ...... day of ...., 19 ...., by and between ...., hereinafter referred to as CONTRACTOR, and ...., hereinafter referred to as SUBCONTRACTOR:

WITNESSETH, that for and in consideration of the payment or payments hereinafter provided to be paid to SUBCONTRACTOR by CONTRACTOR, SUBCONTRACTOR agrees to perform for CONTRACTOR the services hereinafter specified, all under the terms and conditions hereof.

FIRST: SUBCONTRACTOR agrees to ....
....
....

Such work to be performed in .... Addition, according to plans and specifications of ...., Architects. Performance by SUBCONTRACTOR shall be according to terms, conditions and specifications hereinabove or hereinafter provided.

SECOND: SUBCONTRACTOR shall furnish, at SUBCONTRACTOR'S own cost and expense, all equipment, material and labor necessary to the completion of the job specified above, except such as CONTRACTOR is specifically required to furnish under the terms of Paragraph THIRD hereof. All of such materials and equipment shall be of a type and grade satisfactory to CONTRACTOR, unless specifically prescribed elsewhere herein, and all of such labor shall be performed by competent workmen and in a good and workmanlike manner, all in accordance with the plans and specifications which have been made available to SUBCONTRACTOR, and with Veterans Administration and/or Federal Housing Administration Requirements Form No. 2221, Revised March, 1954, including all late revisions.

THIRD: CONTRACTOR shall furnish only the following materials and equipment ....
....
....

Acceptance by SUBCONTRACTOR of such material and equipment shall constitute an agreement by SUBCONTRACTOR that they are fit and proper for the use intended.

FOURTH: SUBCONTRACTOR shall, before the start of performance hereunder, examine any preceding work for line, grade, finish and condition, and shall report any discrepancies to the job superintendent. Beginning of performance shall constitute acceptance of job conditions.

FIFTH: For the services to be performed by SUBCONTRACTOR, hereunder, CONTRACTOR, shall pay the sum of .... to be paid in a lump sum upon final completion and acceptance unless otherwise hereinafter specifically provided.

SIXTH: SUBCONTRACTOR shall perform his entire work or appropriate portion thereof in strict accordance with the schedule of work furnished by CONTRACTOR. Notification of work scheduled to be accomplished by SUBCONTRACTOR will be furnished in writing not later than Friday of the week preceding the week during which the work is scheduled for accomplishment. SUBCONTRACTOR shall diligently prosecute scheduled work to completion and shall complete said work in accordance with the schedule, unavoidable delays due to floods, storms, and other causes beyond the SUBCONTRACTOR'S control excepted.

SEVENTH: If SUBCONTRACTOR fails to prosecute the work with diligence or fails to comply with other requirements herein, CONTRACTOR shall have the right to its election, either through its own employees or through some other contractor, to take over and complete the work. In such event SUBCONTRACTOR shall not be entitled to any payment for work performed by it prior to such taking over until all work provided for herein is completed and accepted by CONTRACTOR, at which time CONTRACTOR'S expenses in completing the work shall be deducted from the amount SUBCONTRACTOR would have received for full performance, and the difference, if any, shall be paid by CONTRACTOR to SUBCONTRACTOR under the conditions herein provided.

EIGHT: SUBCONTRACTOR shall indemnify and save CONTRACTOR harmless from all claims for damage to persons or property caused by SUBCONTRACTOR'S operations hereunder.

FIGURE 3-2

Subcontractor Agreement

NINTH: SUBCONTRACTOR agrees to (1) Comply fully, during the operations performed hereunder, with the Workmen's Compensation laws; (2) Carry Public Liability Insurance; (3) Carry automobile Public Liability Insurance covering all automotive equipment used by SUBCONTRACTOR in performing the obligations of this agreement. SUBCONTRACTOR shall furnish CONTRACTOR evidence of such insurance, showing that said insurance is in full force and effect. Upon SUBCONTRACTOR'S failure to comply strictly with the terms of this paragraph, CONTRACTOR shall have the right, if it so elects, to terminate this agreement.

TENTH: SUBCONTRACTOR agrees to reimburse CONTRACTOR for any loss or damage of any nature whatsoever suffered by CONTRACTOR as the result, either directly or indirectly, of SUBCONTRACTOR'S negligence in operating hereunder. As work progresses SUBCONTRACTOR agrees to clear, pile and remove any debris resulting from SUBCONTRACTOR'S operations. Failure of SUBCONTRACTOR to comply with the above shall entitle CONTRACTOR to have such work done and to charge SUBCONTRACTOR the cost of same.

ELEVENTH: Upon completion of the work to be performed by SUBCONTRACTOR hereunder, and before final payment, SUBCONTRACTOR shall, if requested, furnish CONTRACTOR with proof that all bills for materials, labor and equipment used on the job have been paid, which proof shall include receipted bills for all such items.

TWELFTH: In performing hereunder, SUBCONTRACTOR shall fully comply with all applicable laws, ordinances, rules and regulations of federal, state and local governments.

THIRTEENTH: SUBCONTRACTOR shall at all times occupy the position of an independent contractor, and neither SUBCONTRACTOR nor any of SUBCONTRACTOR'S agents, servants or employees shall ever be considered as the agents, servants or employees of CONTRACTOR.

FOURTEENTH: SUBCONTRACTOR agrees for a period of one (1) year after final payment, all work done hereunder and all materials furnished by SUBCONTRACTOR hereunder against defects, and agrees promptly to repair or replace any and all defective work or materials which may develop within such period, all at SUBCONTRACTOR'S expense. SUBCONTRACTOR shall furnish to CONTRACTOR

a maintenance bond for that period in the amount of $..........

FIFTEENTH: SUBCONTRACTOR'S work shall be subject to the inspection and approval of CONTRACTOR and of representatives of the Federal Housing Administration and/or Veterans Administration, and final payment shall be conditioned upon such approval. Should a reinspection fee be incurred because of defective work or materials, SUBCONTRACTOR agrees to pay the amount of such fee.

SIXTEENTH: This agreement shall be binding upon the heirs, successors or assigns of the parties hereto, but no assignment hereof by SUBCONTRACTOR shall be valid without the prior written consent of CONTRACTOR thereto, nor shall SUBCONTRACTOR sub-contract any of the work to others without such written consent of CONTRACTOR.

EXECUTED at .... the day and year first above written.

ATTEST:

....

....

CONTRACTOR

BY: ....

ATTEST:

....

....

SUBCONTRACTOR

....

BY: ....

FIGURE 3-2 (cont.)

from the sixth paragraph of this contract form: "Subcontractor shall perform his entire work or appropriate portion thereof in strict accordance with the schedule of work furnished by the contractor." It is very important that this schedule of work be made a part of the contract and that sufficient time and emphasis be devoted to this schedule. The importance of this entire sixth paragraph cannot be overemphasized, and certainly a thorough understanding should be had between the contractor and the subcontractor at the time of signing the contract.

Penalties for failure to comply with the schedule are outlined in the seventh clause of the contract. This also should be positively brought to the attention of the subcontractor. Paragraphs eight and nine cover liability and insurance requirements of the subcontractor. It is not sufficient to take this matter of insurance for granted. It must not be taken lightly. The subcontractor must have full insurance for the protection of everyone concerned, and evidence of adequate insurance should be made available to the contractor for his protection. Contractor should require that the subcontractor present the actual policies with complete information pertaining to his coverage, and record should be made of insuring company, policy amounts, numbers, and other pertinent data.

In the eleventh paragraph, the subcontractor is required to "furnish contractor with proof that all bills of materials, labor, and equipment used on the job have been paid, which proof shall include receipted bills for all such items." To have full protection against liens, there are additional steps the general contractor may take. He may obtain a complete list of the suppliers and workers of the subcontractor, together with names, addresses, social security numbers, etc., and follow the procedure used successfully by other contractors. The subcontractors are paid weekly. *All checks are made out jointly to the subcontractor and the supplier or worker.* This method has eliminated the problem of hidden liens for labor or materials at the end of the job. It was developed as the result of a lot of costly experience and has proved very satisfactory in actual operation.

The subcontractor's agreement discussed here includes only the basic elements of the contract. It is frequently necessary to refer specifically to plans and specifications where there is any danger of misinterpretation or discrepancy. The thoroughness of the sub's understanding can only be obtained by discussing the job with him in detail. Addenda to the contract should emphasize and clarify any questionable phases of the job. Reference to this addenda should be made in the first paragraph of the contract. Plans and specifications become an integral part of the contract; although this is assumed by the contractor, it is frequently only partially understood by the subcontractor. Attachment of copies of plans, specifications, and additional clarification is frequently essential to complete communication. All subsequent change orders and extra work orders should definitely become a part of the contract and a copy furnished to the subcontractor. To facilitate the handling of this type of work, Problem Solution Associates has available a Pocket System (small pocket notebook) containing extra work order forms, contract forms, waiver of lien forms, time cards, any many other forms. Description of this Pocket System will be found on page 106.

Occasionally it is to the advantage of the contractor to furnish materials or lend equipment to the subcontractor to expedite the job. If this can be anticipated, it should be made a part of the original contract; however, if this situation develops later, any arrangement for rent or reimbursement for materials should definitely be put in writing, carry the signature of both parties, and become a part of the contract.

Another common source of difficulty is found in unusual conditions met at the job site by the subcontractor. If the subcontractor encounters any unusual conditions on the job which will handicap or interfere with his operations, he should report such conditions immediately to the contractor for proper action.

If the job is to progress in an orderly manner, it is necessary that a continuous clean-up program be maintained. This is covered in the Subcontractor Agreement form discussed here and should be called to the attention of the sub. In some cases, however, the builder agrees to provide his own clean-up labor. In either case, the contractor should have an understanding with the subcontractor regarding clean-up. This should be brought to the attention of all subs bidding so that their bids can be adjusted accordingly. On larger jobs it is usually less expensive for all concerned for the contractor to furnish clean-up labor. This reduces the chance for misunderstanding and friction on the job between the various trades.

The final clause in this Subcontract Agreement covers method of payment. If it is possible for the general contractor to handle weekly progress payments, we advise this method. Since many subcontractors are inadequately financed, this is a definite asset in that capable subcontractors, who would not otherwise be financially able to handle it, may bid on the job. If monthly or weekly progress payments are made, about 10 percent should be retained until final completion of the project.

A primary essential to a smooth contractor-subcontractor relationship is establishment of complete and adequate communications. If the subcontractor is kept aware of the contractor's problems, and vice versa, a closer and more satisfactory working relationship can be established and maintained.

It is recommended that, except on the smallest projects, the general contractor call a meeting to include his own key people, all subcontractors, and suppliers (or representatives). An enthusiastic presentation of the forthcoming project will prove very beneficial to all concerned. An opportunity is afforded for the suppliers and subcontractors to become acquainted, thus improving working relations and reducing chances for friction. This meeting affords an opportunity for everyone to learn exactly what is expected of him and how his part will fit into the overall operation. There will usually be some questions and objections which can be clarified.

Perhaps the best time for this type of meeting is in the evening. The contractor may make it a dinner meeting—but avoid too many drinks! Some contractors have found it beneficial to hold regular meetings, perhaps once a month, during the course of a construction project. This affords an opportunity for everyone to air his problems and invariably results in improved relationships. Contractors who have held such meetings feel that their cost is more than justified by the resulting improved communications and smoother operations.

Since the maintenance of planned progress schedules of the job is the most important single element in profitable operation, the importance of adherence to the schedules by suppliers and subcontractors cannot be overemphasized. Periodic meetings afford an opportunity for ironing out scheduling difficulties and discussing discrepancies in material and workmanship. When these problems are brought up at a meeting, it serves to reduce the danger of repetition among other subcontractors or suppliers. Although it would not be appropriate to place anyone on the spot regarding construction deficiencies and discrepancies, some general examples might be used and individual embarrassment avoided. It is to the general interest of all concerned that deficiencies and discrepancies be held to a minimum, since resulting delays are costly to all.

Each subcontractor must be fully aware of the importance of doing his own part of the job thoroughly and accurately. Otherwise, he leaves costly problems for following trades to combat. The general contractor should emphasize again and again the importance of every subcontractor's leaving his part of the work complete and giving consideration to those trades which follow his. Also each subcontractor must realize his obligation to correct faults or discrepancies discovered later. If he is required to post a maintenance bond covering complaints or callbacks after job completion, he will be more inclined to maintain the necessary quality of work.

Contractors and subcontractors who adhere to these principles will certainly eliminate the vast majority of the undesirable, unnecessary, and expensive friction and problems which arise in the contractor-subcontractor relationship.

WHAT YOU CAN DO TO AVOID DELAYS CAUSED BY CARRIERS

When shipping schedules are close, it will pay to investigate the reliability of carriers between your location and your source. Once I shipped a piece of machinery by motor freight from Memphis to San Diego. The carrier people assured me that they went to San Diego. What they failed to tell me was that they went by way of Kansas City and Salt Lake City. The shipment was two weeks longer than anticipated arrival and caused a costly delay. Needless to say, I learned to check on routings. If delivery is critical, you may investigate routes and carriers and select the best. Then the carrier can request his office at the point of origin to follow through and make the pick up. This provides an additional means of follow-up without cost and often expedites delivery.

The expediter or person assigned this job should fully realize the importance of this function and never get so busy with other functions that he neglects his expediting duties. These must come first if projects are to stay on schedule. Too often expediting is assigned to someone who already has too much to do.

FORMS AND PROCEDURES FOR HANDLING BIDS

Abstract of Bids

Two forms for listing of bids are shown in Figures 3-3 and 3-4. Both of these are government forms. They are shown primarily to give you ideas for forms of this type.

INTERNAL PLANNING AND CONTROL OF THE PROJECT 59

FIGURE 3-3

Abstract of Bids Opened

The printer can adjust or rearrange the items, adding or omitting any that are not desired. Or, as previously mentioned, the ideas can be used as a starter to assist in laying out the desired forms on the drawing board. Methods of forms preparation are discussed in detail in Chapter 6, Streamlined Office Procedures and Forms.

ABSTRACT OF BIDS OR INFORMAL PROPOSALS

SHEET NO.

FIRM NAMES ▶											
ACCEPT-ANCE TIME											
PAYMENT DISCOUNT											
DELIVERY TIME											
F.O.B. POINT											
ITEM NO.											
PURCHASE ORDER NO.											

(Check all bids carefully prior to award.)

THIS IS TO CERTIFY that the above informal proposals were secured by the undersigned.

DATE INFORMAL PROPOSALS SECURED	PROPOSALS SECURED BY ☐ PERSONAL VISIT ☐ TELEPHONE ☐ WRITTEN PROPOSAL
SIGNATURE	TITLE

THIS IS TO CERTIFY that the above bids were secured by formal advertisement under the Invitation No. shown below.

INVITATION NO.	DATE BIDS OPENED	HOUR BIDS OPENED
SIGNATURE - RECORDER	TITLE	
SIGNATURE - BID OPENING OFFICER	TITLE	

VA FORM MAR 1969 07-2232 EXISTING STOCKS OF VA FORM 07-2232, MAR 1962, WILL BE USED. *(OVER)*

NAMES AND ADDRESSES OF FIRMS INVITED TO BID OR SUBMIT INFORMAL WRITTEN PROPOSALS AND PUBLIC BUILDINGS WHERE POSTERS WERE DISPLAYED

(Not required for informal solicitation)

FIRMS INVITED TO BID

BUILDINGS WHERE POSTERS WERE DISPLAYED

EXPLANATION OF OTHER THAN LOWEST BID (PURCHASES), HIGHEST BID (SALES)

EXPLANATION

ATTESTATION OF RESULTS OF DRAWING OF TIE BIDS

NAME OF BIDDER DRAWN

SIGNATURE OF EMPLOYEE WHO DREW SLIP	SIGNATURE OF CONTRACTING OFFICER

WITNESSES

NAME	ADDRESS
NAME	ADDRESS
NAME	ADDRESS

FIGURE 3-4

Abstract of Bids or Informal Proposals

INTERNAL PLANNING AND CONTROL OF THE PROJECT

RECORD OF BIDDERS

Project _____ No. _____ Date _____
Location _____ Class of Work _____
Date Released for Bids _____ Date Bids Due _____

BIDDERS NAME ADDRESS AND PHONE	PLANS		SPECS		DEPOSITS		ADDENDA AND REMARKS
	OUT	IN	OUT	IN	IN	RET'D.	

PS 72002

FIGURE 3-5

Record of Bidders

CONFIRMATION OF TELEPHONE QUOTATION

TO

FIRM

ADDRESS

JOB

ADDRESS

DATE

ESTIMATE NUMBER

PHONE

JOB NO.

DESCRIPTION OF WORK	AMOUNT OF BID
	TOTAL
	TAX
	TOTAL BID

EXCLUSIONS AND REMARKS

QUOTATION PREPARED BY

PS 72016

FIGURE 3-6

Confirmation of Telephone Quotation

(Example 1)

INTERNAL PLANNING AND CONTROL OF THE PROJECT 63

CONFIRMATION OF TELEPHONE QUOTATION

TO: _____ DATE _____

FIRM _____ JOB NO. _____

ADDRESS _____

JOB ADDRESS AND OWNER, IF DIFFERENT _____

THIS IS TO CONFIRM OUR VERBAL QUOTATION TO YOU OF _____ CONCERNING THE FOLLOWING:

DESCRIPTION	QUOTATION
TAX, IF ANY	
TOTAL	

QUALIFICATIONS AND EXCLUSIONS, IF ANY

THANK YOU FOR THE OPPORTUNITY TO SUBMIT THIS QUOTATION.

SUBMITTED BY _____

P S 72006

FIGURE 3-7

Confirmation of Telephone Quotation

(Example 2)

64 INTERNAL PLANNING AND CONTROL OF THE PROJECT

Record of Bidders

It should go without saying that a record of all bidders should be kept. The form shown in Figure 3-5 provides the essential information. This function should be handled by one person, or at least placed under the control of one person so that all information is recorded as it takes place.

Confirmation of Telephone Quotations

Two sample forms, Figures 3-6 and 3-7, provide a method of writing out and mailing information given over the telephone. Without the use of such a form, quotations given over the telephone can become a source of trouble by misunderstandings and misinterpretations. The bid should be put in writing immediately after the telephone conversation.

Shop Drawing Approval

Many costly errors can be circumvented by giving careful attention to shop drawings made by others. It is not sufficient to take a casual glance at the drawing and say, "Looks O.K. Go ahead." A suggested design and wording for an approval stamp is shown in Figure 3-8. After carefully checking the shop drawing, use this stamp to

☐ APPROVED   ☐ APPROVED AS CORRECTED
☐ NOT APPROVED ☐ REVISE AND RESUBMIT

CHECKING IS ONLY FOR CONFORMANCE WITH THE DESIGN CONCEPT OF THE PROJECT AND COMPLIANCE WITH THE INFORMATION GIVEN IN THE CONTRACT DOCUMENTS. CONTRACTOR IS RESPONSIBLE FOR THE DIMENSIONS TO BE CONFIRMED AT THE JOB SITE, FOR INFORMATION THAT PERTAINS SOLELY TO THE FABRICATION PROCESSES AND/OR TECHNIQUES OF CONSTRUCTION, AND FOR COORDINATION OF THE WORK OF ALL TRADES.

_____   _____
BY   DATE

FIGURE 3-8

Shop Drawing Approval Stamp

INTERNAL PLANNING AND CONTROL OF THE PROJECT 65

indicate the approval, or other action given. By all means, avoid verbal orders or commitments in this connection. Your local office supplier or printer can have the stamp made.

Progress Reports

The form shown in Figure 3-9 is a "Purchase and Hire Progress Report" on which a summary of the project status is recorded. Although this is a VA form, the general idea is good for top management information used in financial planning and control. This is presented as an idea starter for the design of a form to accomplish the individual requirements.

Construction Cost Estimate Forms

Most estimators have their own system of recording the components of their estimates. Ledger pads or other ruled sheets are customarily used. Small jobs may be estimated on sheets such as the VA form shown in Figure 3-10. As most contractors already know, the Frank R. Walker Company specializes in forms for contractors; for more on this, see page 191.

Joint Venture Agreements

As construction projects become larger and more complex, joint ventures are becoming more common. The following contract form (Figure 3-11) will provide a good start toward reducing the agreement to writing. It will provide a checklist of important points to be covered. Additional specific points may be added to the agreement as desired in each case.

One additional area which may be included is the placing of responsibility for complying with health and safety laws. Each contractor furnishing equipment should bear the responsibility for compliance of the equipment with all safety laws.

PURCHASE AND HIRE PROGRESS REPORT

TO		PROJECT NUMBER		REPORT FOR PERIOD ENDING
		PROJECT LOCATION		

1. DESCRIPTION OF PROJECT

2. DATE CONSTRUCTION STARTED		8. DATE PHYSICALLY COMPLETED	
3. SCHEDULED % OF PHYSICAL COMPLETION		9. AVERAGE WORK FORCE PER DAY IN PERIOD	
4. ACTUAL % OF PHYSICAL COMPLETION		10. MAN-HOURS WORKED THIS PERIOD	
5. EST. VALUE OF WORK PLACED IN PERIOD	$	11. VALUE OF WAGES EARNED THIS PERIOD	$
6. SCHEDULED COMPLETION DATE		12. INITIAL WORKING COST ESTIMATE	$
7. ESTIMATED COMPLETION DATE			

LINE NO.	BRANCH OF WORK (A)	PRESENT WORKING COST ESTIMATE		EST. VALUE OF WORK IN PLACE TO DATE	
		MATERIALS (B)	LABOR (C)	MATERIALS (D)	LABOR (E)
13					
14					
15					
16					
17					
18					
19					
20					
21					
22					
23					
24					
25					
26					
27					
28					
29					
30					
31					
32					
33					
34					
35					
36					
37					
38					
39					
40					
41					
42					
43					
44					
45					
46					
47	TOTAL (Lines 13 thru 46)	$	$	$	$
48	TOTAL (Materials and labor)		$		$
49	ESTIMATED COST OF SURPLUS MATERIALS INCORPORATED INTO PROJECT TO DATE			$	
50	ESTIMATED COST OF STATION LABOR TO DATE			$	

VA FORM OCT 1971 **08-6090** SUPERSEDES VA FORM 08-6090, NOV 1966, WHICH WILL NOT BE USED. 497161

FIGURE 3-9

Purchase and Hire Progress Report

51	VALUE OF CONTRACTS AWARDED AND PURCHASE ORDERS ISSUED TO DATE	$
52	VALUE OF WAGES EARNED BY PURCHASE AND HIRE EMPLOYEES TO DATE	$
53	TOTAL FUNDS OBLIGATED AND COMMITTED TO DATE *(Add lines 51 and 52)*	$
54	CONTRACT AND PURCHASE ORDER PAYMENTS APPROVED TO DATE	$
55	TOTAL ACCRUED EXPENDITURES TO DATE *(Add lines 52 and 54)*	$

FUNDS ALLOTTED FOR THE PROJECT

LINE NO.	APPROPRIATION SYMBOL (A)	ALLOTMENT RESTRICTION (B)	DATE ALLOTTED (C)	AMOUNT (D)
56				$
57				
58				
59				
60				
61				
62				
63				
64	TOTAL FUNDS ALLOTTED TO DATE *(Add lines 56 thru 63)*			$
65	UNOBLIGATED BALANCE *(Line 64 minus line 53)*			
66	FUNDS REQUIRED TO COMPLETE *(Line 48, Column C, minus line 64)*			

DESIGN AND CONSTRUCTION SCHEDULING *(Omit after construction starts.)*

67. DESIGN DATA		SCHEDULE APPROVED	SCHEDULE EST./ACT.	68. CONSTRUCTION DATA	SCHEDULE APPROVED	SCHEDULE EST./ACT.
PRELIMINARY PLANS*	A. START DATE			A. ISSUE INVITATION TO BID*		
	B. % COMPLETE			B. BIDS OPEN*		
	C. COMPL. DATE			C. CONSTR. START *(P&H or Contr. Award)*		
WORKING DWGS. & SPECS.*	D. START DATE			D. COMPLETE CONSTRUCTION		
	E. % COMPLETE			69. NAME AND ADDRESS OF ARCHITECT-ENGINEER*		
	F. COMPL. DATE					

*When applicable to station supervised project.

70. EMPLOYEES PAID FROM 36×0108 ACCT. 9900 *(Clerical Assistants)*			71. P&H EMPLOYEES PAID FROM 36×0108 PROJECT PAYROLL	
1A.	1B.	1C.	2A.	2B.

REMARKS *(Special progress and/or delaying factors)*

72. SIGNATURE OF RESIDENT ENGINEER

73. SIGNATURE OF ENGINEER OFFICER *(If designated as Resident Engineer)*

FIGURE 3-9 (cont.)

CONSTRUCTION COST ESTIMATE

DATE PREPARED

SHEET NO. OF

PROJECT

LOCATION

ARCHITECT-ENGINEER

BASIS FOR ESTIMATE

☐ CODE A *(No design completed)*
☐ CODE B *(Preliminary design)*
☐ CODE C *(Final design)*
☐ OTHER *(Specify)*

DRAWING NO.

ESTIMATOR

CHECKED BY

SUMMARY	QUANTITY		LABOR		MATERIAL		TOTAL COST
	NO. UNITS	UNIT MEAS.	PER UNIT	TOTAL	PER UNIT	TOTAL	

VA FORM SEP 1964 **08-6238**

FIGURE 3-10

Construction Cost Estimate

INTERNAL PLANNING AND CONTROL OF THE PROJECT 69

JOINT VENTURE AGREEMENT: SAMPLE FORM

THIS AGREEMENT MADE THIS ___ DAY OF _____, BY AND BETWEEN (NAME OF FIRST FIRM) , A (PARTNERSHIP, CORP, ETC.) HAVING ITS PRINCIPAL PLACE OF BUSINESS AT (STREET) (CITY) (STATE) , HEREINAFTER SOMETIMES REFERRED TO AS _____ AND (NAME OF SECOND FIRM) , A (PARTNERSHIP, CORP, ETC,) HAVING ITS PRINCIPAL PLACE OF BUSINESS AT (STREET)   (CITY)   (STATE) HEREINAFTER SOMETIMES REFERRED TO AS _____.

WITNESSETH:

WHEREAS, THE PARTIES HERETO HAVE BEEN AWARDED THE CONTRACT FOR THE _____ _____. OF THE (NAME OF PROJECT) IN (LOCATION) SUCH CONTRACT BEING HEREINAFTER REFERRED TO AS THE "CONTRACT"; AND,

WHEREAS, THE PARTIES DESIRE THAT THEIR INTERESTS IN THE SERVICES TO BE RENDERED AND THE WORK TO BE DONE UNDER THE CONTRACT AND ANY PROFITS DERIVED THEREFROM, AND ANY LIABILITY FOR LOSSES ARISING OUT OF THE PERFORMANCE THEREOF, BE DEFINED BY AN AGREEMENT IN WRITING;

NOW, THEREFORE, FOR AND IN CONSIDERATION OF THE PREMISES AND THE MUTUAL UNDERTAKINGS HEREINAFTER SET FORTH, THE PARTIES HEREBY CONSTITUTE THEMSELVES AS JOINT VENTURERS SOLELY FOR THE PURPOSE OF PERFORMING THE SERVICES NECESSARY FOR COMPLETION OF THE CONTRACT AND TO CARRY OUT THEIR JOINT VENTURE, THE PARTIES HERETO DO AGREE AS FOLLOWS:

1. NATURE OF JOINT VENTURE. THE CONTRACT SHALL BE ENTERED INTO IN THE NAMES OF THE PARTIES AS JOINT VENTURERS, AND THE JOINT VENTURE SHALL BE KNOWN AS THE (NAME OF PROJECT) JOINT VENTURE.

2. SHARE OF PROFITS AND LOSSES. THE INTERESTS OF (1ST FIRM) AND (2ND FIRM) IN AND TO THE CONTRACT, AND IN AND TO ANY AND ALL MONEYS WHICH MAY BE DERIVED FROM THE PERFORMANCE THEREOF AND THE OBLIGATIONS AND LIABILITIES OF EACH OF THE PARTIES HERETO AS AMONG THEMSELVES IN CONNECTION WITH THE CONTRACT AND WITH RESPECT TO ANY AND ALL LIABILITIES AND LOSSES IN CONNECTION THEREWITH, SHALL BE EQUAL (EXCEPT AS PROVIDED WITH RESPECT TO DIVISION OF PROFITS IN THE EVENT EITHER PARTY DOES NOT FURNISH ITS PROPORTIONATE SHARE OF WORKING CAPITAL AS PROVIDED IN PARAGRAPH 3). EACH PARTY DOES HEREBY INDEMNIFY THE OTHER AGAINST ANY LOSS OR LIABILITY EXCEEDING THE PROPORTIONS HEREINABOVE STATED BY REASON OF ANY LIABILITY INCURRED OR LOSS SUSTAINED IN AND ABOUT THE CONTRACT.

3. WORKING CAPITAL

(A) ALL NECESSARY WORKING CAPITAL, WHEN AND AS REQUIRED FOR THE PROSECUTION OF THE CONTRACT, SHALL BE FURNISHED BY THE PARTIES EQUALLY. A BANK ACCOUNT SHALL BE OPENED AT THE (NAME OF BANK) , OR SUCH OTHER BANK AS THE PARTIES MAY DETERMINE, IN WHICH ALL THE FUNDS RECEIVED ON ACCOUNT THEREOF, SHALL BE DEPOSITED. WITHDRAWALS SHALL BE MADE UPON CHECKS SIGNED BY THE REPRESENTATIVES OF EACH OF THE PARTY FIRMS.

(B) IT IS CONTEMPLATED THAT WORKING CAPITAL REQUIREMENTS WILL BE PROVIDED BY BORROWING ON THE PART OF THE JOINT VENTURE FROM THE (NAME OF BANK) , SUCH BORROWINGS TO BE SECURED BY A PLEDGE OF THE CONTRACT. THE PARTIES AGREE WITHIN FIFTEEN (15) DAYS FOLLOWING EXECUTION OF THIS AGREEMENT TO EXECUTE SUCH NOTES, INSTRUMENTS OR ASSIGNMENT, AND OTHER DOCUMENTS AS MAY BE NECESSARY OR APPROPRIATE FOR THE ARRANGING OF SUCH FINANCING. ALL NOTES EVIDENCING BORROWING BY THE JOINT VENTURE SHALL BE SIGNED BY A REPRESENTATIVE OF EACH OF THE MEMBER FIRMS.

THE NEED FOR WORKING CAPITAL IN ADDITION TO ANY FUNDS AVAILABLE FROM SUCH FINANCING BY (NAME OF BANK) SHALL BE DEPOSITED BY EACH PARTY AND SHALL BE DETERMINED AS FOLLOWS:

FIGURE 3-11

Sample Contract Form

WITHIN FIFTEEN (15) DAYS AFTER EITHER OF THE PARTIES DETERMINES THAT ANY ADDITIONAL SUMS ARE REQUIRED FOR THE PERFORMANCE OF THE CONTRACT, BOTH PARTIES SHALL DEPOSIT IN SUCH BANK ACCOUNT SUCH EQUAL AMOUNTS AS SHALL BE DESIGNATED BY THE PARTY MAKING THE DETERMINATION.

(C) IN CASE EITHER PARTY IS UNABLE OR FAILS OR NEGLECTS TO CONTRIBUTE ITS PROPORTIONATE SHARE OF THE WORKING CAPITAL REQUIRED TO PERFORM THE CONTRACT, THEN THE OTHER PARTY MAY, BUT SHALL NOT BE REQUIRED TO, ADVANCE SUCH DEFICIENCY OR ANY PART THEREOF. IF THE OTHER PARTY DOES ADVANCE SUCH SUM, THAT PARTY SHALL BE ENTITLED TO RECEIVE INTEREST COMPUTED AT THE RATE OF __% PER ANNUM ON THE EXCESS ADVANCED OVER THE PERIOD OF TIME THAT SUCH EXCESS CONTRIBUTION EXISTS. THE AMOUNT OF SUCH INTEREST SHALL NOT BE TREATED AS AN EXPENSE OF THE JOINT VENTURE, BUT SHALL BE CHARGED AGAINST THE ACCOUNT OF THE PARTY FAILING TO ADVANCE ITS SHARE OF WORKING CAPITAL.

(D) ALL WORKING CAPITAL ADVANCED SHALL BE REPAID TO THE PARTY ADVANCING THE SAME PRIOR TO THE DISTRIBUTION OF ANY PROFITS HEREUNDER. NO PART OF ANY WORKING CAPITAL ADVANCED TO THE JOINT VENTURE SHALL BE RETURNED TO EITHER PARTY AND NO DISTRIBUTION OF PROFIT SHALL BE MADE PRIOR TO THE COMPLETION OF THE CONTRACT EXCEPT AS MAY OTHERWISE BE MUTUALLY AGREED UPON IN WRITING BY THE PARTIES.

4. CREDIT TO OTHER PARTY. EXCEPT AS PROVIDED HEREIN, NEITHER PARTY SHALL HAVE THE RIGHT TO BORROW MONEY ON BEHALF OF THE OTHER PARTY, NOR TO USE THE CREDIT OF THE OTHER PARTY FOR ANY PURPOSE.

5. SELECTION OF REPRESENTATIVES. TO FACILITATE THE HANDLING OF ALL MATTERS AND QUESTIONS IN CONNECTION WITH THE PERFORMANCE OF THE CONTRACT, EACH OF THE PARTIES SHALL APPOINT REPRESENTATIVES TO ACT FOR IT IN ALL SUCH MATTERS, WITH FULL AND COMPLETE AUTHORITY TO ACT ON ITS BEHALF IN RELATION TO ANY MATTERS OR THINGS IN CONNECTION WITH, ARISING OUT OF, OR RELATIVE TO, THE JOINT VENTURE AND IN RELATION TO ANY MATTERS OR THINGS INVOLVING THE PERFORMANCE OF THE CONTRACT, INCLUDING, BUT NOT LIMITED TO, THOSE OF A CONTRACTUAL NATURE WITH _____(OWNER)_____ AND/OR __(HIS, ITS)_____ REPRESENTATIVES, OR WITH THIRD PERSONS. EACH PARTY SHALL NOTIFY THE OTHER IN WRITING OF THE NAME OF ITS REPRESENTATIVE TO ACT UNDER THE PROVISIONS OF THIS PARAGRAPH AS SOON AFTER THE EXECUTION OF THIS CONTRACT AS PRACTICABLE. EITHER PARTY MAY AT ANY TIME AND FROM TIME TO TIME CHANGE ITS REPRESENTATIVE BY FILING WITH THE OTHER A NOTICE IN WRITING OF SUCH SUBSTITUTION, BUT UNTIL THE APPOINTMENT AND FILING OF SUCH NOTICE, THE ACTIONS OF THE REPRESENTATIVE OR ALTERNATE HEREBY APPOINTED SHALL BE CONCLUSIVELY BINDING UPON SUCH PARTY.

6. ACTIONS OF REPRESENTATIVES. THE REPRESENTATIVES OF THE PARTIES SHALL MEET FROM TIME TO TIME AS REQUIRED TO ACT ON NECESSARY MATTERS PERTAINING TO THE CONTRACT. ALL DECISIONS, COMMITMENTS, AGREEMENTS, UNDERTAKINGS, UNDERSTANDINGS OR OTHER MATTERS PERTAINING TO THE PERFORMANCE OF THE CONTRACT SHALL BE MUTUALLY AGREED UPON BY SUCH REPRESENTATIVES. WHEREVER PRACTICAL ALL CONFERENCES OF A SUBSTANTIAL NATURE WITH THIRD PARTIES SHALL BE ATTENDED BY THE REPRESENTATIVES OF BOTH FIRMS. RECORDS OF ALL CONFERENCES SHALL BE KEPT AND COPIES FURNISHED TO THE RESPECTIVE PARTIES.

7. PUBLICITY. ALL NEWS RELEASES RELATING TO THE _(NAME OF PROJECT)_ AND THE PARTIES SHALL BE SUBJECT TO APPROVAL BY THE PARTIES. IN REFERRING TO ANY ARCHITECTURAL SERVICES OR ARCHITECTS IN CONNECTION WITH THE _(PROJECT)_, THE NAMES OF THE TWO FIRMS SHALL BE EMPLOYED JOINTLY RATHER THAN THE NAMES OF ANY INDIVIDUALS OF SUCH FIRMS.

8. RESPONSIBILITIES OF THE PARTIES. WHILE EACH OF THE PARTIES HERETO RECOGNIZES THAT IT IS JOINTLY AND SEVERALLY RESPONSIBLE WITH THE OTHER PARTY FOR THE PERFORMANCE OF ALL OF THE OBLIGATIONS UNDER THE CONTRACT, THE PARTIES ALLOCATE CERTAIN OF THE

FIGURE 3-11 (cont.)

RESPONSIBILITIES AND OBLIGATIONS BETWEEN THEMSELVES AS FOLLOWS:

(A) __(1ST FIRM)__ SHALL BE RESPONSIBLE FOR _____

_____

(B) A JOB COORDINATOR WILL BE FURNISHED BY __(1ST FIRM)__, WHO WILL BE RESPONSIBLE FOR COORDINATION OR WORK BETWEEN THE TWO FIRMS, AND BETWEEN THE FIRMS AND __(OWNER)__.

(C) __(2ND FIRM)__ SHALL BE RESPONSIBLE FOR _____

_____

(D) ENGINEERS WILL BE SELECTED BY MUTUAL AGREEMENT.

(E) IMMEDIATELY FOLLOWING THE EXECUTION OF THIS AGREEMENT, THE PARTIES SHALL PREPARE A PROGRAM FOR THE ACCOMPLISHMENT OF THE SERVICES CALLED FOR UNDER THE TERMS OF THE CONTRACT WHICH SHALL BE EMPLOYED TO FACILITATE AND COORDINATE THE EFFORTS OF THE PARTIES, BUT WHICH SHALL NOT BE TREATED AS A BINDING CONTRACT BETWEEN THEM.

9. EXPENSES. EACH PARTY SHALL BE REIMBURSED ON THE FIRST DAY OF EACH MONTH FOR SUCH PARTY'S EXPENSES INCURRED IN BEHALF OF THE JOINT VENTURE DURING THE PRECEDING MONTH AT THE RATES OF REIMBURSAL SHOWN BELOW.

(A) ALL DIRECT COSTS SHALL BE PAID OUT OF THE JOINT VENTURE ACCOUNT, INCLUDING, BUT NOT LIMITED TO TYPING, PRINTING AND REPRODUCTION, ENGINEERS FEES, MODEL AND/OR RENDERINGS, PHOTOGRAPHY, LEGAL AND ACCOUNTING EXPENSES, TRAVEL, SEPARATE PROFESSIONAL LIABILITY INSURANCE, AND ANY OTHER COSTS DIRECTLY RELATED TO THE JOB.

(B) THE PARTIES SHALL BE ENTITLED TO BE REIMBURSED FOR SERVICES PERFORMED ON BEHALF OF THE JOINT VENTURE BY THE PRINCIPALS OR EMPLOYEES OF THE PARTIES ON THE BASIS OF ACTUAL TIME SPENT BY SUCH INDIVIDUALS IN BEHALF OF THE JOINT VENTURE. REIMBURSEMENT SHALL BE UPON THE FOLLOWING SCALE.

(1) PRINCIPALS ACTING AS COORDINATORS OR IN A GENERAL EXECUTIVE CAPACITY—$_____ PER HOUR.

(2) PRINCIPALS ACTING AS DESIGNERS, SPECIFICATION WRITERS, OR SUPERVISORS—$_____ PER HOUR.

(3) PRINCIPALS ACTING AS DRAFTSMEN – A RATE EQUAL TO THE HIGHEST RATE PAID ANY DRAFTSMEN ASSIGNED TO THE JOB.

(4) REIMBURSEMENT FOR ALL OTHER EMPLOYEES SHALL BE AT THEIR CUSTOMARY HOURLY RATE OF COMPENSATION AS EXISTING AS OF THE DATE HEREOF.

(C) IN ADDITION TO REIMBURSEMENT AT THE RATES SET FORTH ABOVE, THE PARTIES SHALL BE ENTITLED TO BE REIMBURSED AT THE RATE OF $_____ PER HOUR IN RESPECT OF OVERHEAD FOR EACH HOUR CHARGED TO THE JOB AS DEFINED IN THE PRECEDING SUBPARAGRAPH.

10. BOOKS AND RECORDS.

(A) SEPARATE BOOKS OF ACCOUNT FOR THE PERFORMANCE OF THE CONTRACT IN ALL MATTERS THERETO PERTAINING SHALL BE KEPT AND MAINTAINED AT THE OFFICES OF __(1ST FIRM)__, WHICH RECORDS AND BOOKS SHALL BE OPEN FOR INSPECTION OF THE PARTIES AT ALL REASONABLE TIMES.

(B) MAINTENANCE OF SUCH BOOKS AND RECORDS SHALL BE UNDER THE SUPERVISION OF __(ACCOUNTING FIRM)__, INDEPENDENT CERTIFIED PUBLIC ACCOUNTANTS.

(C) UPON THE COMPLETION OF THE CONTRACT A TRUE AND CORRECT ACCOUNTING SHALL BE HAD OF ALL COSTS AND EXPENSES, AND ALL ACCOUNT VOUCHERS RECORDS AND DATA RELATING TO THE CONTRACT AND ITS PERFORMANCE.

(D) TO THE EXTENT THAT THE RECORDS OF THE JOINT VENTURE MUST BE SUBSEQUENT TO THE COMPLETION OF THE CONTRACT, PURSUANT TO THE PROVISIONS OF THE LAW, THE SAME

FIGURE 3-11 (cont.)

shall be kept at such place or places as the parties may from time to time determine, and the cost thereof shall be borne equally by them.

11. <u>Division of profits.</u> Upon the completion of the project, after providing for and paying (a) all costs disbursed or incurred in the performance of the Contract; (b) all other costs and charges ordinarily and usually charged as costs in the performance of such a contract; (c) any and all claims not secured by insurance; (d) proper reserves for any claims which shall either have been brought against the parties or which the parties may reasonably anticipate will be brought against them; and (e) reserves for contingencies, if any, that shall be determined by the parties in their discretion to be necessary, and after repaying all borrowings of the Joint Venture, and all sums advanced by the parties for working capital, any profits thereafter remaining resulting from the performance of the Contract, shall be distributed and divided equally between __(1st firm)__ and __(2nd firm)__. Any reserves, when no longer required, or so much thereof as shall remain, shall be similarly distributed.

12. <u>Burden of losses.</u> (a) If the performance of the Contract results in a loss, the parties shall be obligated equally for any such loss. Such equal liability of the parties for the bearing of losses shall continue with respect to any claims which at any time, either before or after the completion of the Contract, shall be made against them, or either of them by reason of this Joint Venture or any matter or thing in connection therewith.

(b) In the event of loss:

(1) If any funds remain, and both parties have contributed their required proportions of working capital, then such remaining funds shall be paid to the parties in the amounts contributed by each, less their respective shares of the loss.

(2) If both parties have not contributed their required proportions of working capital, but sufficient funds are available, then such funds shall be repaid to the parties in the amount contributed by each, less their respective shares of the loss.

(3) If both parties have not contributed their required proportions of working capital, and there are insufficient funds to accomplish the division prescribed in the preceding subdivision, and if there is a deficit in the account of one of the parties by reason of its failure to contribute its required proportion of working capital, then such defaulting party shall make up the deficit in its account. Upon its failure to do so, the indemnity provisions of paragraph 2 of this agreement shall become operative, so as to insure that the non-defaulting party shall bear no more than its proportionate share of the loss.

(4) If both parties have not contributed their required proportions of working capital, and no funds remain or some liabilities are unsatisfied, then the indemnity provisions of paragraph 2 of this agreement shall become operative, so as to insure that neither party shall bear more than its proportionate share of the loss.

13. <u>Insolvency.</u> In the event of the bankruptcy or insolvency of either party, or in the event either party commits any act of bankruptcy or takes advantage of any bankruptcy, reorganization, composition, or arrangement statute, then, from and after such date, such party (hereinafter referred to as the "insolvent party") and its representative and alternate, as hereinbefore referred to (anything in this agreement to the contrary notwithstanding), shall cease to have any voice in the management of the Joint Venture. All acts, consents, and decisions with respect to the performance of the work under the Contract shall thereafter be taken solely by the other party, its representative and alternate. Notwithstanding the foregoing, the insolvent party

FIGURE 3-11 (cont.)

INTERNAL PLANNING AND CONTROL OF THE PROJECT

SHALL REMAIN LIABLE FOR ITS SHARE OF LOSSES, AND SHALL BE ENTITLED TO RECEIVE ITS SHARE OF ANY PROFITS, SUCH PROFITS TO BE PAID AT THE TIME AND IN THE MANNER PROVIDED BY THIS AGREEMENT.

14. <u>LIMITS OF JOINT VENTURE.</u> THE RELATIONSHIP BETWEEN THE PARTIES SHALL BE LIMITED TO THE PERFORMANCE OF THE CONTRACT IN ACCORDANCE WITH THE TERMS OF THIS AGREEMENT. THIS AGREEMENT SHALL BE CONSTRUED AND DEEMED TO BE A JOINT VENTURE FOR THE SOLE PURPOSE OF CARRYING OUT THE CONTRACT. NOTHING HEREIN SHALL BE CONSTRUED TO CREATE A GENERAL PARTNERSHIP BETWEEN THE PARTIES, NOR TO AUTHORIZE EITHER PARTY TO ACT AS GENERAL AGENT FOR THE OTHER PARTY, NOR TO PERMIT EITHER PARTY TO BID FOR OR TO UNDERTAKE ANY OTHER CONTRACTS FOR THE OTHER PARTY.

15. <u>ASSIGNMENT.</u> NEITHER THIS AGREEMENT NOR ANY INTEREST OF EITHER OF THE PARTIES HEREIN (INCLUDING ANY INTEREST IN MONEYS BELONGING TO OR WHICH MAY ACCRUE TO THE JOINT VENTURE IN CONNECTION WITH THE CONTRACT, OR ANY INTEREST IN THE JOINT ACCOUNTS, OR IN ANY PROPERTY OF ANY KIND EMPLOYED OR USED IN CONNECTION WITH THE CONTRACT) MAY BE ASSIGNED, PLEDGED, TRANSFERRED OR HYPOTHECATED, WITHOUT THE PRIOR WRITTEN CONSENT OF THE PARTIES HERETO.

16. <u>TRUST FUNDS.</u> ALL MONEYS CONTRIBUTED BY THE PARTIES TO THIS JOINT VENTURE AND ALL MONEYS RECEIVED AS PAYMENTS UNDER THE CONTRACT OR OTHERWISE RECEIVED SHALL BE TREATED AND REGARDED AS, AND ARE HEREBY DECLARED TO BE, TRUST FUNDS FOR THE PERFORMANCE OF THE CONTRACT, AND FOR NO OTHER PURPOSE UNTIL THE CONTRACT SHALL HAVE BEEN FULLY COMPLETED AND ACCEPTED BY THE OWNER, AND UNTIL ALL OBLIGATIONS OF THE PARTIES HERETO SHALL HAVE BEEN PAID, OR OTHERWISE DISCHARGED, OR PROVIDED FOR BY ADEQUATE RESERVES. SUCH RESERVES SHALL LIKEWISE BE TREATED AS TRUST FUNDS UNTIL THEY HAVE SERVED THE PURPOSE FOR WHICH THEY WERE CREATED.

17. <u>ARBITRATION.</u> ALL DISPUTES WITH RESPECT TO THE PERFORMANCE OF THE CONTRACT SHALL BE SUBMITTED TO __(1ST ARBITRATOR)__ AND __(2ND ARBITRATOR)__ AS ARBITRATORS WHO SHALL SELECT A THIRD ARBITRATOR, THE DETERMINATION OF A MAJORITY OF WHOM SHALL BE FINAL.

IN WITNESS WHEREOF, THE PARTIES HERETO HAVE CAUSED THESE PRESENTS TO BE SIGNED THIS THE DAY AND DATE FIRST ABOVE WRITTEN.

_____(1ST FIRM)_____

BY _____

_____(2ND FIRM)_____

BY _____

FIGURE 3-11 (cont.)

4

Utilizing the Computer to Reduce the Cost of Estimating, Scheduling and Designing

Recently the executive secretary of one of the California state bureaus made news when he discarded his million-and-a-half-dollar computer in favor of hand operations. In this instance, he claims he was able to reduce his staff from 240 people to 106, and at the same time reduce the processing time from 95 days to 10 days. Computers definitely have valuable applications, but they are no panacea for all ills. Some of the salesmen for computer time and other proponents of computer applications are over-enthusiastic about their value. Each application must be analyzed by the contractor's own management to weigh the anticipated costs and advantages against present methods.

ESTIMATING BY COMPUTER

It is usually best to enter into computer use by the time-sharing method. Then, if your operations are sufficiently extensive to justify your own setup, a full installation may be considered. If your estimating is extensive and you have a real need for fast estimates, then consider the computer. One of the pioneers of computer estimating is Allen Brothers and O'Hara, Memphis, Tennessee, who became so interested in the potential of computer estimating that they formed a separate organization for this purpose and obtained their own computer. Since their own work did not keep the computer busy, they sold time to others. They do not actually perform the estimating for others, but rather provide a method, unit prices, and the computer facility time.

The main advantage in computer estimating is that estimates can be prepared in a matter of a few hours compared with several days for the average project. In using the computer for the preparation of cost estimates, the same basic procedures are followed as for manual estimating. First, a quantity take-off is made from plans and specifications. Then the measurements are listed on a format sheet for the computer. The computer is programmed with unit costs so that calculations can then be made. This method eliminates many long hours of slow calculations and numerous sources of errors. The important factor here is to provide the computer with accurate unit costs based upon actual historical records of similar work. Estimator's opinions are minimized. Separate printouts can be obtained for subcontractors, vendors, labor costs, and financial requirements. Also, quantity surveys and man-hour requirements are obtained for the use of scheduling, whether CPM or other methods. Almost any statistical information can be obtained.

COMPUTER APPLICATION TO SCHEDULING

CPM is a natural extension of computer estimating, since the computer already has most of the information needed to print out the schedule. CPM scheduling is really not radically different from the usual procedures. Essentially, the numbering system is the same as that recommended by the Associated General Contractors except that two additional digits are used to further break down and identify the exact segments of the classification. Quantity take-offs are listed on format sheets for units of construction instead of by various trades. Sizes of crews are determined and the computer multiplies this by the already stored labor costs to arrive at the time required for each activity. This information is then placed on the CPM arrow diagram. Most of the problems associated with accurate scheduling by the CPM have been resolved, and the system is a definitely proved scientific method of scheduling and project control.

There are now several computer programming organizations with considerable experience in construction scheduling, estimating, and other computer applications, and it is best to profit from their experience in getting started with computer applications.

SUBCONTRACTORS AND THE COMPUTER

The general contractor can run off a copy of the CPM diagram and listing of the various subdivisions of the work for the subcontractors. This information will provide starting and completion dates as well as the relationship of the various trade activities. It will show crew sizes which have been used for estimating purposes. Most subcontractors find CPM a big advantage because it lets them know just where they stand. Too often, subcontractors have had to work almost in the dark. Is it any wonder they have developed a casual attitude about scheduling? CPM provides the subcontractor with the information he needs, and must have, if he is to properly organize his own operations.

The subcontractor whose job is complex can also benefit by detailing his own operations and developing his own CPM schedule. This can be especially valuable in scheduling work such as mechanical or electrical jobs on large projects. The subcontractor's CPM can be made to coincide with the general CPM for the entire project, but obviously cannot be made final until the overall schedule has been established. CPM is applied equally well to engineering type projects such as roads, dams, pipelines, and others.

ADVANTAGES OF THE COMPUTER IN DESIGNING

Based on historical cost data, a preliminary estimate of a proposed project can be quickly prepared by the computer. Then, if the cost figure thus obtained is higher than desired, a new approach can be made to the design. For example, if reinforced concrete construction has been tentatively planned, other less costly construction may then be considered. Or a less expensive design may be the alternative. In any event, design work need not proceed to completion before cost data is obtained, as is usually the case. Many projects have been scrapped entirely because owners have been "sold" a particular design or construction only to find too late that the project cannot be built within the budget.

Also, after the project is in a more advanced stage, another estimate can be run to determine whether design is still within budget, or if some trimming is necessary. Relative costs are quickly available on the numerous types of construction and the many surface treatments available. Obviously, it is far better to make changes while the design work is in process than to wait until it has been completed.

As the design work progresses, a more accurate estimate evolves since more and more of the variables have become firm. If the computer process is utilized, there are few surprises at the completion of design work.

COMPUTER DRAFTING

Traditionally, all drafting and drawing has been a laborious, slow and tedious task requiring the application of man-hours. Now for the first time in history, drawings can be made by computerized machinery. Although computer drafting cannot replace the draftsman, it certainly can reduce by a wide margin the long back-breaking hours necessary to produce a completed set of drawings for a construction project. The programming language is relatively simple and a draftsman can learn to program the special computer in a short time.

A side benefit of computer drafting is its inherent accuracy. Rate of drawing is about 200 inches per minute or more. Capabilities of the computerized drafting machine are rapidly expanding. It can be programmed to draw a perspective in three dimensions. A complete floor plan for a building of several thousand square feet can be drawn in ten to fifteen minutes—a job which would require several days of a draftsman's time. By taking the drudgery out of drafting, the job becomes not only more attractive, but actually fascinating. Wiring diagrams, plumbing and mechanical systems drawings drawn by the hand method are nothing but hard work. This is where the drafting machine really shines. It never complains, never asks for a raise, is seldom sick, and is not highly emotional. Just punch the right information into a paper tape, fill the ink pen, sit back on the high stool and watch 'er go!

One such drafting is manufactured by Kongsberg Vaapenfabrikk of Norway. Dietzgen Corporation, with offices in most principal cities, has the Orthograph Coordinate Measuring and Digitizing machine with full complement of automatic drafting equipment. The method is not yet well adapted to architectural drafting, but progress is being made.

A complete set of equipment may cost in excess of $100,000 at the present time. As with other electronic equipment, costs can be remarkably reduced as production increases.

ENGINEERING APPLICATIONS OF THE COMPUTER

The computer enables the engineer to collect, store, and retrieve the tremendous amount of data with which he works. Obviously, this vastly increases the functions he can perform and greatly reduces the time required. He can now collect and organize for future use all of the information pertinent to his projects. The potential is limited only to the engineer's ingenuity. Subdivision of land, for instance, normally requires much detail work which can be greatly simplified by the computer. High-voltage power transmission lines can be engineered by the computer, including location of towers, spacing, height, strength and costs. Optimum location over irregular terrain can be established by computer processing of the necessary information. The most economical structural design and materials can be determined by computer. Even factors of maintenance, operation and convenience can be programmed.

Vast experience in these and other areas of computer applications are available from several of the service organizations such as IBM, NCR, and others. For additional information and sources of information on computer applications, see Chapter 12.

THE ROLE OF THE COMPUTER IN MANAGEMENT

Computers can assist management in many ways, and provide the following advantages:

1. The capability to produce almost instant answers to questions and problems requiring computations and statistical analysis of changing conditions.
2. The ability to store and retrieve almost limitless amounts of information and relate this information to other data.
3. The ability to rapidly compute the various alternative possibilities associated with innumerable facts and conditions as an aid in making management decisions.
4. The ability to forecast the results of various management proposals and provide the information needed to determine the most satisfactory.

In effect, the computer can make available to management more information, more accurately and faster, than is possible by any other method.

5

How Visual Control Systems Can Assist in Communicating Your Ideas

For adequate control we must have *measurement, reporting, comparison,* and *analysis* of essential business information. First, we measure performance. Then we report this information so that it can be compared with existing standards and analyzed. If we are to visualize and convey this information to others we must have some method of graphical representation; to be successful in this communication effort the visual presentation must be clear, of adequate size, well designed, and flexible enough to allow changes and revisions as required.

VISUAL AIDS

There are many types of visual aids, among which are graphs, charts, display boards, three-diminsional models, photos, drawings, and others. New materials and techniques developed in recent years make it possible for a person of average talent to produce professional quality charts and visual presentations right in his own office.

Pressure-Sensitive Transfers

One of these aids is known as "pressure-sensitive transfers." Various sizes and styles of alphabets, numerals, symbols, cross-hatching, arrows, dotted lines and other designs are printed on a clear plastic film and then transferred to your layout simply by rubbing the top side of the transfer sheet. A number of manufacturers have catalogs of extensive lines of these pressure-sensitive transfers. A list of sources will be found in Chapter 12 under the heading "Sources of Supply for Charts, Graphs, and Forms."

CPM and other planning displays can be made rapidly using these pressure-sensitive transfers. The dense black inking facilitates the making of copies by Xerox, offset printing, or other methods of reproduction. Various colors are also available in many of the items. Lines and border designs come on plastic rolls which are simply stuck on the surface like Scotch tape. Work flow symbols can be used to make plant layouts showing flow of materials and work in process. Data processing symbols may be used to show the flow of forms and paper work through the business offices. The uses are almost unlimited. Also see "You Can Design and Lay Out Your Own Business Forms" in Chapter 6.

Laminated Plastic Sheets

For making display boards, a laminated plastic sheet material is available. These are white with light blue grids of various sizes. The boards are reusable and can be easily laid out or changed as desired. They are ideal for making office or plant layouts. Plan views of all types of furniture, work benches, machines and equipment are available as models or stick-on items. They are simply attached to the grid to facilitate working out floor plans. These aids permit quick and simple repositioning to attain the desired results.

Magnetic Visual Control Systems

A system using magnetic symbols and a steel wall chart makes a very flexible display board. Magnetic accessories include letters, numerals, arrows, card holders, and symbols in a great variety. These are simply placed on the board where desired and can be easily changed as required. Magnetic symbols of various types of construction equipment can be used to set up equipment schedule boards or for maintenance schedules.

Other applications for such boards include organizational charts, various types of scheduling, personnel organization, vacation schedules, machine loadings, sales, dispatch, apartment status, and many others. Catalogs and information on this type of system are available from Methods Research Corporation, 105 Willow Ave., Staten Island, N.Y. 10305, or Executive Planning, Inc., Asbury Park Road, Farmingdale, N.J. 07727.

A wall planning chart showing an entire year's activities has been designed for many applications. A special kit of components has been designed for making network diagrams such as CPM. Others include floor plan layouts, vehicle maintenance, job progress, accident statistics and many others. This system is produced by Pryor Marking Products, 21 East Hubbard St., Chicago, Ill. 60611.

HOW COSTS CAN BE LOWERED BY METHODS ENGINEERING

Methods engineering techniques have been used for many years by manufacturers to lower operating costs. More recently these techniques, along with new advancements in equipment, make it possible for the contractor to affect substantial cost reductions on projects of every type.

The original tools used by the methods engineer were stopwatch and clipboard. These are not well suited to construction methods study, in most instances, due to the fact that construction activities are usually spread over a wide area and comprise numerous busy operations. For these busy and complex activities, time lapse photography or video tape recorders with instant playback capability is far superior.

Time Lapse Photos

Motion picture equipment with single frame feature is used to take photos at regular intervals. Timing may be from one frame per second to one every two, three, or four seconds. This will provide a good record of job activities which may be viewed and studied for ways to improve operations. Even on routine jobs, a systematic study by a qualified methods engineer using time lapse photography or video tape can almost invariably lower the costs of construction. Without time lapse equipment or a video tape recorder the methods engineer is wasting 90 percent of his *own time*. Isn't this contrary to the objectives of methods engineering? It is much like digging trenches by hand compared to modern trenching equipment.

Time lapse photos may be taken on 8mm, Super 8mm, or 16mm film. Although larger sizes may be used, there would seldom be any advantage in using anything over 16mm. Personally, I much prefer the 16mm frame size since it is four times the area of 8mm and about three times that of Super 8mm. The obvious advantage is that you get three or four times the detail. Since the camera location may be 200 to 300 feet from the scene, important details may be lost in anything smaller than the 16mm size film. My own time lapse equipment consists of a 16mm Bolex 16 SBM camera fitted with a Vario-Switar 100 POE zoom lens and essential timing mechanism. This lens provides continuous zooming action, electrically driven, for covering the desired amount of the scene from a single vantage point.

Although workmen rarely object to the time lapse study, it is best to remain as inconspicuous as practical. If the camera is kept a considerable distance from the scene, workers are seldom aware that photos are being taken. If they notice the activity, they soon ignore it as they go about their work. My own equipment is automatic and requires little or no attention for hours at a time. I usually set up the camera equipment and start the time lapse photos before the men arrive on the job. In fact, I have done undercover work, photographing the action from inside my Airstream trailer. Since the windows are double glazed, tinted and curved, activities inside the trailer can not be observed from outside. Windows on the sides, upper contours and ceiling of the trailer allow good visibility of the entire scene from excavation to high rise. This is especially valuable in the filming of activities on higher floors.

Various intervals of time may be used between frames, but the range of one frame per second to one frame every two, three, or four seconds is normally used. There are some actions requiring faster sequence. Even normal movie speed is sometimes used. One frame every three or four seconds usually is sufficient to get the necessary information for the study; at this rate, a half-day of operations can be viewed and discussed in an hour or two.

Color film is used since it is no more costly than black-and-white and affords vastly superior identification of men and activities. Processing can now be done overnight in most average size cities. Film sent out of the area by air mail can reach its destination and be processed within 24 hours.

One of the most important requirements for a successful time lapse system is the resolving power of the lens. For good picture quality, a high-grade lens is a must. One of the problems with most 8mm equipment is the relatively poor quality of the lens. A good telephoto lens with built-in exposure control is also very important. The zoom lens permits exact coverage of the amount of activity desired without the necessity for revolving a turret or repositioning the equipment. It is important to hold the activity to a minimum so that work crews are not distracted. The less commotion created the better.

The contractor may wish to purchase his own equipment and assign someone to make the studies; however, a qualified methods engineer or consultant who specializes in this type of work may be able to accomplish more and at less expense. Cost to purchase a complete setup of Super 8mm equipment is about $2,000; for 16mm it is about $3,000. Qualified methods engineers charge around $300 per day. Some charge by the roll of film. In any case you get about what you are willing to pay for. A good methods engineer taking 16mm time lapse photos and several still shots per day can usually effect savings amounting to several times his cost. Don't forget that when methods are improved, this same improvement will be reflected in lower job costs every time that operation is done in the future.

After the film is returned from the processor it is studied, preferably on a Moviscop or editor, and notes are written or dictated concerning the findings. It is amazing how often slow, unproductive techniques become visible for the first time when viewed frame by frame where the action can be analyzed. No individual can take in the entire swarm of activities when viewing them on the scene, but when photos are studied later, everything becomes clearly visible. Time studies become meaningful by this method and can result in faster, more efficient operations and consequently lower job costs. Safety is improved simultaneously.

For individual study of time lapse film, I use a Zeiss Moviscop 16mm editor. I view every one of the 4,000 frames in each 100 foot roll individually, and tabulate the results on a chart which presents a glaring, vivid portrayal of the losses invariably present on the average construction operation. Concurrently with the discovery of lost time and motion, I make a safety survey which results in safer operations. Safety is not only desirable because it is required by law (see Chapter 10, Occupational Safety and Health); safer operations also amount to lower operating costs due to lower insurance rates, less time lost on the job, and improved schedules.

After the film is thoroughly analyzed and every effort made to locate lost time and devise improved methods, a conference of the operating management of the project is called. The film is then presented on a 16mm projector especially designed for analyzing the film (see "Methods Engineering Services and Equipment," Chapter 12). With it the film can be run at normal speed, or reduced in speed down to single frame viewing without burning the film. The ordinary projector does not have this capability.

Remember also that ordinary movie equipment is not capable of taking time lapse photos. The camera should have an electric drive, otherwise it must be hand wound frequently. It must have a single frame capability, and this operation must be timed for exposures at regular intervals. The timing mechanism alone may cost as much or more than the camera. Considerable time, money and frustration can be avoided by consulting someone with experience in time lapse. Few camera and photo equipment suppliers have anyone knowledgeable on this highly specialized subject. For further information and for recommendations on equipment, see "Methods Engineering Services and Equipment" in Chapter 12.

Video Tape Recorders and Monitors

This type of equipment has the advantage of instant playback, but it has several disadvantages as well. First, the equipment is larger and more difficult to place on the construction project. It requires 115 v. current, and the fidelity, or detail, in the picture is not comparable to the 16mm photos. Enlargements of 16mm photos can be made when desired for purposes of instruction or other visual presentations. Video tape does not offer this capability.

Another important advantage of film is that it will be in color. Video recordings will not. Color affords an excellent means of identification of workmen. In charting the information from the film, it is necessary to identify each worker by his hard hat or clothing. In the black-and-white video this can not be done as well.

The advantage of instant playback, however, can not be equaled in any other way. This can be especially valuable when devising new construction methods since changes in the job procedures can be inaugurated at once if desired. Cost of the equipment is now within the cost range of a good set of 16mm equipment, that is, about $2,500 to $3,000, assuming 115 volt current is available. Tapes may be retained if desired, although the cost is several times that of 16mm film. See "Methods Engineering Services and Equipment" in Chapter 12.

6

Streamlining Your Office Procedures and Forms to Reduce Errors and Save Time

Among the infinite variety of office forms available, some of the most useful have been selected and included in this chapter. Those forms which originate in the field are shown in the appropriate chapter. Accounting and bookkeeping forms are not included since these are usually selected by the accountant.

JOB COST DISTRIBUTION

One of the problems facing the bookkeepers is to properly distribute costs among the various jobs. Accurate distribution is essential if job costs are to be determined accurately. If job costs are not accurately determined, the contractor is working in the dark. He might as well guess at costs if his cost records on previous jobs are not precise. Bookkeepers can perform this cost distribution only when they are provided with the right information.

When invoices come in on materials used on more than one job, someone must provide the correct quantities used on the various jobs. If some of the materials went into the warehouse, this must be indicated on the paperwork. In order to simplify this function a rubber stamp such as the one shown in Figure 6-1 may be used to stamp the statements and invoices. Provision is made for a breakdown and distribution to three jobs, but if more than three allocations are to be made, simply make two or more impressions.

```
               JOB COST DISTRIBUTION
               BY ―――――― DATE ――――――
               JOB NO.  | JOB NO. | JOB NO.
               ─────────┼─────────┼─────────
               $        | $       | $
```

FIGURE 6-1

Job Cost Distribution Stamp

THE INTANGIBLE BENEFITS OF COLOR CODING

The value of color coding has been proven many times, yet most firms are not taking advantage of this simple device to help prevent errors, save time and facilitate filing and record keeping. As many of you know, the Problem Solution Procedure Manual for contractors is color coded as follows:

Subject	*Color Code*
Organization	Buff
Finance	Lt. Gray
Personnel	Green
Office Practice	Pink
Legal	Dark Blue
Insurance	Dark Gray
Accounting	Brown
Tax	Yellow
Communications	Beige
Equipment	Salmon
Collections	Golden

Subject	Color Code
Materials	Lt. Green
Methods	Taupe
Estimating	Lt. Brown
Project Control	Lt. Blue
Alert Bulletins	Cherry Red

This may at first seem to be a lot of unnecessary work, but it has really been much more than worth the trouble. The color code facilitates the filing of additional sheets or revision material in the exact place. Anything not properly filed is valueless. Do you have any idea how much time is lost in the average organization searching for misfiled or mislaid papers? When forms and paperwork are printed on a specific color paper the problem of misfiling is minimized and a form of a certain type is easily spotted.

Your printer can obtain paper of any color from among the dozens of colors available. Many printers resist using specific colors because they may have to order them. The stock is not far away, so insist on getting what you want. We use seventeen distinct colors, plus white, and it is no problem to us.

YOU CAN DESIGN AND LAY OUT YOUR OWN BUSINESS FORMS

Standard business forms are seldom exactly right for your needs. With a drawing board and a few simple supplies you can prepare your own layouts for forms. In the preparation of the black and white copy used for reproduction, various line drawings or illustrations are "pasted up" and photographed. Then an offset printing plate is made from this photo. Only the black ink of the illustration is picked up by the camera and the outlines of pasted-on parts do not appear on the final print job. Here is a list of essentials for this work:

1. A drawing board of some type. The draftman's large table and drafting machine are ideal but not essential. A small drawing board, T-square and triangle will do the job.
2. An ink ruling pen and ink compass for circles. A complete drafting set is nice to have, but not necessary for most work.
3. A good grade of paper, 100 percent rag content with a smooth surface, will be needed.
4. Forms such as the one shown in Figure 3-6 can be made with only a typewriter. The type shown is upper and lower caps on an IBM Executive electric typewriter. This typewriter uses carbon plastic tape instead of the more conventional fabric ribbon. Be sure to use black ink on all layout work. Never use blue, especially the lighter shades, since it will not be picked up by the camera.
5. For lettering of a different size or character than your typewriter characters, a product known as "pressure-sensitive transfer type" is available from art supply and drafting supply firms. It is used in advertising and graphic arts work to eliminate the tedious and slow hand-lettering operation. These transfers are simple to use. Just place the desired character in the spot to be used and rub the back of the plastic sheet on which it is mounted. It transfers firmly to your layout sheet. An almost endless variety of

these designs is now available. I use these transfers in making up layouts for business forms and illustrations for this and other books. Among the various designs available are many styles of lettering, numerals, arrows, decorative borders, screens and symbols.

6. Sharp, clean lines can be obtained by using transfer tape. It comes on a roll with the dense black (or color if desired) line on plastic tape. If you can handle the ruling pen satisfactorily, it will have a broader application since it is infinitely variable in width. Use only black waterproof drawing ink—never writing ink. Lines can be made with a black ball point pen, but they will not be as sharp. Obviously, only one width of line is available with the ball point pen, whereas the ruling pen is capable of making a line of any desired width.

With the above tools and supplies, a professional job of forms layout can be done. Just be careful and keep the layout neat. If you have not had much experience in using ink, some practice may be required. I made many of the forms in this book and I am not a professional artist.

HOW TO RECOVER A LETTER AFTER IT IS MAILED

Did you know that you can recall a letter after it is mailed? There are times when this knowledge can be very important. The form shown in Figure 6-2 is the correct Post Office form to be used; it may be obtained from almost all post offices. Ask for PS Form 1509.

COMMUNICATIONS FORMS

Some simple forms which will provide written records of essential communications are those shown in Figures 6-3 and 6-4. Both of these forms are designed to fit the ring binders of the Frank R. Walker Company and Problem Solution Associates (see page 191). They are simple, self-explanatory forms which may be used in the office or out in the field.

STREAMLINING YOUR OFFICE PROCEDURES AND FORMS 89

FIGURE 6-2

Sender's Application for Recall of Mail

MEMORANDUM
OF CALL

TO ____________________

YOU WERE CALLED BY ☐ VISITED BY ☐

NAME ____________________

COMPANY ____________________

☐ PLEASE CALL
☐ IS WAITING TO SEE YOU
☐ RETURNED YOUR CALL
☐ WILL CALL AGAIN ____________________ DATE TIME
☐ REQUESTS APPOINTMENT ____________________ DATE TIME

MESSAGE

RECEIVED BY	DATE	TIME

REMARKS

P 2001

FIGURE 6-3

Memorandum of Call

a.v.o.
Avoid Verbal Orders

TO ____________________

FROM ____________________

DATE ____________ JOB NO. ____________

PLEASE SEE THAT THE FOLLOWING IS ACCOMPLISHED.

HAS ALL NECESSARY INFORMATION BEEN INCLUDED?

✓CHECK:	WHEN?	WHY?
WHO?	WHERE?	HOW MANY?
WHAT	HOW?	HOW MUCH?

P2005

FIGURE 6-4

A. V. O.—Avoid Verbal Orders

7

Developing Tight Job-Site Project Control

Too often the superintendent allows ommissions and errors to creep in because he has not given sufficient study to the details of the project. Just recently, I was looking over the plans on a project when I noticed an error in the concrete slab. When I brought this to the attention of the superintendent he said, "Oh well, we can always cut out the concrete." But to "cut out the concrete" required a half-day and equipment which was not on the job.

We all make mistakes, but it seems to me that it would be easier if we applied more effort toward *avoiding* them than to *correcting* them. It is essential, therefore, for the superintendent to make a thorough study of the plans and specifications so that he has a good mental picture of the entire job. To accomplish this, he must devote sufficient time to the study under conditions free from distractions and interruptions.

ANTICIPATING AND PREVENTING MATERIALS AND SUPPLIES PROBLEMS

Obviously, it would not be possible to maintain a schedule unless the correct materials were on hand when needed. In Chapter 8, which deals with materials, you will find a checklist of important points which must be investigated upon arrival of materials on the job. Most delays due to materials problems can be prevented by following this checklist. It is a good idea to have the architect or owner's representative confirm colors, patterns, etc., as soon as possible after arrival on the job. Sometimes colors and patterns vary so much that they are not acceptable, but workmen will naturally install what they are given. The problem arises when part of the lot is several months or a year old, and is mixed with a new shipment. A critical inspection of the entire shipment is the only way to detect such variations before causing delays.

Don't rely on the labels. I have seen incorrect packaging and labeling. Open packages to make positive confirmation of the contents. As I have said, *assume nothing*.

BE ALERT TO DEVIATIONS AWAY FROM SCHEDULE

Regardless of how well the job is planned prior to moving on site, day-to-day conditions will require a constant awareness of any drift away from schedule. The project superintendent, or manager, should study the progress of the job daily so that any trend away from schedule can be seen and analyzed early. A delay in any activity should be analyzed immediately to determine how it will affect the schedule of the project. It is certainly not uncommon to have some seemingly unimportant delay result in an overall project delay later.

Most contractors have set up a daily reporting system. Suggested reporting forms are included in this chapter. If you need something different, your printer can make the changes; or you can lay out the desired forms yourself on the drafting board, in ink, and the printer can make a photo-plate for offset printing. By doing it this way you can be sure of getting exactly what you need. If too much detail makes it difficult to lay out, make it one and one-half or two times larger. It can then be reduced by the printer to the exact size required.

The point is that some system of providing absolutely current information on job progress and job changes is necessary if delays in schedule are to be avoided. Obviously, to be of value this information must be scrutinized carefully and corrective action taken at once.

Let's take the modern manufacturing assembly line and compare it to construction. If management of the manufacturing operation overlooked one screw or one minute part, assembly lines would come to a screeching halt. In construction work, the very small items seldom have such an obvious crippling effect, but they can often disrupt the operations. These disruptions are considered normal in construction activities, but *should they be considered normal?* They are not taken for granted in manufacturing. Why should they be so considered in construction? Believe me, the construction industry as a whole is the most inefficient of all industries, but it does not have to be so. The industry has made considerable improvement in management and operating efficiency in recent years, but there are still plenty of areas of possible cost reduction.

WHAT YOU CAN DO TO KEEP THE PROJECT ON SCHEDULE

It is often difficult for suppliers and subcontractors to fit the project into their schedule because of conflicting commitments. Inevitably, there are going to be delays somewhere along the line, but our objective is to try to anticipate these and take preventive steps when possible. There *are* actions which may be taken to alleviate this annoying problem. On page 123 see "How to Avoid Delays Due to Materials and Supplies Problems" with its accompanying checklist.

DEVELOPING TIGHT JOB-SITE PROJECT CONTROL

This brings us to one of the big advantages of CPM. Without a good scheduling system, suppliers and subcontractors are left in the dark. The contractor tells them that they must be on the job by a certain time, but that usually means very little, because their experience has shown them that the job probably won't be ready for them on time. So how do we now convince them of the accuracy and seriousness of our scheduling?

Go over your schedule with subs and suppliers. A big wall chart is effectively impressive as evidence of the seriousness of your scheduling. Those subs and suppliers whose operations are critical should be made especially aware of the importance of keeping on schedule. Also, it is well to point out the consequences of their failure to handle their part of the responsibilities. An impressive, well-planned graphical representation of the complete project schedule is convincing evidence of your capability in scheduling.

Several visual control systems, such as wall charts and display boards, are described in Chapter 5.

PROVIDE FULL INFORMATION TO THE WORK CREWS

To avoid loss of time and errors on the job, give proper consideration to the information the crews need to keep the job moving. By providing them with construction details that are complete and accurate you enable them to progress faster and with less effort.

The interior room finish schedule shown in Figure 7-1 is an example of the type of information needed to proceed with the finishing-out of interiors of buildings. This form (No. 401) available from the Frank R. Walker Company, is one of the many forms they have for contractors. Whenever information must be conveyed to workmen, it helps to put this into some type of graphical form. A drawing or sketch may be all that is needed, or this may be supplemented with an explanation, checklist, or some other written information. In any case, communications are usually more complete when the proper communications form is used. If standard forms are available, it is less expensive to use them; but where they are not available, consider making up a sketch of the desired form and having some printed.

ATTITUDES ARE CONTAGIOUS

Many construction men do not seem to be aware of the extent to which their own actions and attitudes affect others around them. A study of the methods and techniques of the most successful construction men reveals some interesting facts. After visiting and working around construction men for thirty years, I find the difference between an efficient, smooth-running project and a haphazard operation easily and quickly recognizable. Have you ever noticed these differences? Show me a construction superintendent or foreman who is serious minded, businesslike, and well organized, and I will show you a smooth-running project. Attitudes are contagious. If management men present a disciplined and serious attitude, other men on the job are much more inclined to view their own responsibilities in a similar manner. On the other hand, when the

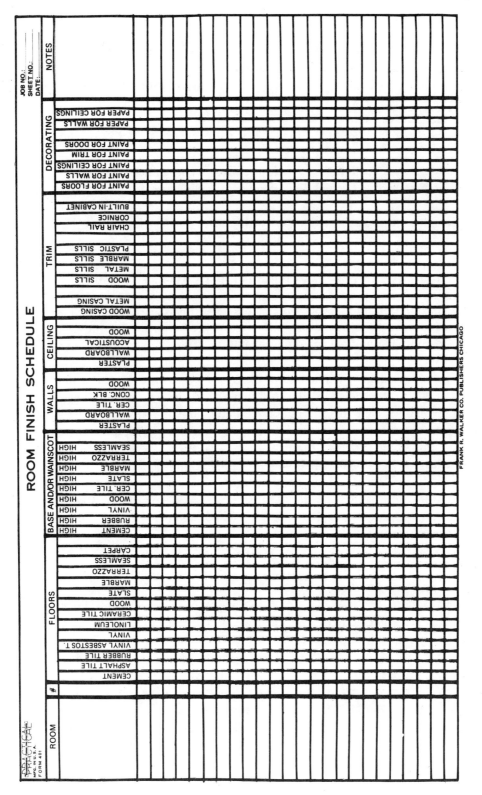

FIGURE 7-1

Room Finish Schedule

(Frank R. Walker Co., Publishers
5030 North Harlem Ave., Chicago, Ill. 60656)

DEVELOPING TIGHT JOB-SITE PROJECT CONTROL 95

boss conducts himself like a practical joker, the entire project may turn out to be one big joke. Any apparent lack of concern on the part of management will certainly be reflected in the attitudes of workers. Consequently, all management men while on the job should conduct themselves in a serious, purposeful manner.

HOW TO PREVENT ACCUMULATION OF PROBLEMS

The checkout of some projects is a regular nightmare due to the huge accumulation of sloppy and incomplete work. And even worse, much of it does not get on the punch list, but requires numerous call-backs later. Call-backs can be time consuming, irritating, and expensive to both owner and contractor. Let's see what can be done to reduce them.

1. Close supervision of your own forces will reveal small problems, incomplete work, untested systems and components, and other potential sources of accumulated trouble. See that corrections are made as the work progresses; otherwise crews move off, taking the tools and materials necessary to do the work. These small items may require ten times the amount of time if overlooked until the end of the project.

2. Subcontractor's operations should be observed to see that his forces have adequate supervision; otherwise, his workmen may be left alone and allowed to drift off the desired course. Keep your eyes open for potential problems. Once I was checking on resilient tile flooring when I noticed that the first line of tile did not look parallel with the wall. Upon checking the measurements, the workman found that he had made a two-inch error. If I had not observed the flaw, he would have had most of the floor down before detecting the error. I saw one floor that was installed with just such a condition and never corrected. The owner did not notice the condition until after the job was accepted. He was so disgusted, he installed carpet over the tile.

3. Components which are to be installed late in the job, such as hardware, should be inspected carefully long before needed. On a school job recently, cabinets were delivered to the job two weeks before installation, but no one noticed that they were a foot short. This delayed the job two weeks. Then, door checks were found to be defective, causing an additional two days' delay. Both these delays could have been prevented by making a careful inspection at the earliest stages.

4. On purchases involving shop drawings, more lead time is necessary to prevent delays. The subcontractor or vendor must prepare the drawings and submit them for approval. Time must be allowed for checking the drawings. Suppose the drawings need revision; additional time will be required. Therefore, see that shop drawings are in for approval in time to be revised if necessary.

5. On items which are being manufactured for the project, such as cabinet or millwork, special electrical boxes, etc., it is important to make actual visits to the shops to see that the work is in progress at the proper time. A check on the quality of the workmanship and materials can often prevent delays due to receipt of goods of unacceptable quality. Of course, this is of even greater importance when dealing with untried vendors.

6. Take immediate corrective action on deliveries not in full accordance with purchase orders, since shortages and damaged items can delay the project.

WORK ORDERS, CHANGE ORDERS AND EXTRA WORK ORDERS

A form to be used either as a Work Order or as a Change Order is shown in Figure 7-2. A convenient checklist is printed along the right hand side to serve as a reminder of the various activities which may be required in conjunction with the order. The contractor's name, address and logo may be imprinted in the upper left hand corner. In addition to the contractor's copy shown, there may be copies for:

1. Owner or customer
2. Project Manager or Superintendent
3. Sales or other file copies
4. Field or foreman

An Extra Work Order form (P2004) fits the Pocket System and is ideal for use in the field where small changes in the contracted work are required. A sample of this form is shown in Figure 7-3, along with several others designed for field use.

Many contractors are still losing too much of their profits through lax procedures of handling changes and extra work items. To prevent this type of loss, definite, closely controlled procedures must be established, with proper forms to be used *without exception*. Also, I should add, *without delay*. Get something in writing *before* the work is done. Any delay can cause complications. The rule here is the same as in selling anything: *Always have a full understanding of price and complete description of service to be performed before the transaction is made.*

HOW TO STAY OUT OF HOT WATER

I think every construction man will know what I am talking about when I say "hot water." I mean trouble, trouble, trouble. There are times when it seems that *nothing* is going to go right. Like bananas, problems come in bunches. We have no magic wand to offer to make them go away, but there are ways to protect yourself and, as I say, to stay out of hot water.

First, equip yourself with a loose-leaf notebook you can carry around with you, if you don't have one already. Keep a diary of the activities of each day. Some may not seem important at the time, but may develop into a problem later. I can cite numerous instances where I would have been in plenty of hot water if it hadn't been for the detailed diary my wife kept day by day. A record of the exact date and time of day a certain event occurred can be a life saver to you. The most difficult part is developing the habit of keeping the diary, but once you get into the habit, it will be like grabbing a cigarette is for those smokers who have the habit.

Another important method of record keeping is the camera. Any camera will do. I like to use the small lightweight 35mm. Some construction people use the Polaroid camera because they know at once whether they have a picture or not. This is a matter of individual preference and is unimportant. The important thing is to get photos of all accidents, improper procedures, violations of any rules by anyone, whether your own forces or others. Job status photos may be important in some later controversy.

DEVELOPING TIGHT JOB-SITE PROJECT CONTROL 97

☐ **WORK ORDER**
☐ **CHANGE ORDER** No..............

REFER:
Job No..............
Sheet No..............
Of..............

Customer's Name..............
Address..............
Job Location..............
Date..............
Date Promised..............
Job Phone..............
Authorized by:

Description of Work Ordered/Changed:

✓ LIST
Start
Permit
Water & Power
Matl's. Ordered
Subs Ordered
Tools and Equipment
Owner Notified
Keys?
Date Completed
Materials Picked Up
Equipment Picked Up
Work Accepted
Billing Data Audited

PS 72003   CUSTOMER'S COPY

FIGURE 7-2

Work Order—Change Order

```
                    EXTRA WORK ORDER

No. _____  Date _____  19 ___
Job _____
To _____
Address _____
Please furnish all Materials and Labor necessary to complete the following
work, and charge to our account as noted below:




                                            



The work covered by this order shall be performed under the same Terms and
Conditions as that included in the Original Contract.
 1. The above work to be paid for at actual cost of Labor and Materials, plus
    _____ percent (_____%).
 2. All of the above work to be completed for the sum of_____
    _____Dollars.

Signed _____
P 2004
```

FIGURE 7-3

Extra Work Order

Such records may settle a dispute favorably for you when nothing else could. It is a human failing to transpose the sequence of events. The only way you can prove your contentions is with sufficient records. If the photo will obviously be important later, consider including a calendar page plainly showing the date the photo was taken. This is hard to argue with later, since there is no chance that you have forgotten when the photo was taken.

DEVELOPING TIGHT JOB-SITE PROJECT CONTROL 99

As the Chinese say, a picture is worth ten thousand words. I know contractors and subcontractors who take a series of photos during the progress of the work. One heating contractor sets up a camera focused on the work crew and automatically takes a sequence photo every ten minutes. He says this once helped him achieve a favorable court decision because he had photos to disprove the claims of his customer.

This method of taking sequence photos at regular intervals is known as time lapse photography. An interesting discussion of the great potentials of this technique appears in Chapter 5, Visual Control Systems.

Daily Construction Report Forms

A daily construction and material report form (shown in Figure 7-4) provides considerable information on the daily progress of the work. This is Form 110-111, 9¼″ x 11⅞″, available from the Frank R. Walker Company, Publishers, 5030 North Harlem Avenue, Chicago, Ill. 60656. Daily Job Report No. PS 72007, shown in Figure 7-5, is a flexible form which can be adapted to many types of small jobs. These are available from Problem Solution Associates (see page 211).

Other daily construction report forms are described on page 106. These forms are small pocket-size fillers for pocket ring binders described on page 106.

Labor Report Forms

A simple form for reporting labor time on specific jobs is shown in Figure 7-6. Labor for the entire week goes on this form. This is Form No. PS 72004, a Problem Solution form. A similar type form is shown in Figure 7-7a, (Form No. PS 72001), also a Problem Solution form. It is a job record sheet providing information on employees' time in the shop and on the job. The reverse side of this sheet (shown in Figure 7-7b) provides detailed cost records on the job.

PRACTICAL
FORM 111
MFD. IN U. S. A.

DAILY MATERIAL REPORT

JOB NO: 103 DATE June 6, 19--
NAME OF WORK Smith Store Building WEATHER Clear
LOCATION 656 N. Main St., Arlington, Ill. TEMPERATURE 62 7 A.M. 83 2 P.M.

RAILROAD AND CAR NO.	CLASS OF WORK	MATERIALS RECEIVED AND FROM WHOM	QUANTITY
	Cem. Floors	Ready-mixed concrete - Arlington Ready-Mix Co.	9 cy
	Masonry	Common brick - Arlington Builders Supply Co.	8000
	Rough Carpentry	62 Pcs. - 2	
		31 " - 2	
		25 " - 1	

SUMM

KIND OF MATERIAL	Concrete Sand	Gravel	Crushed Stone
PREVIOUS			
TO-DAY			
TOTAL			
KIND OF MATERIAL	**Fire Brick**	**Special Brick**	**Reinforcing Steel**
PREVIOUS			
TO-DAY			
TOTAL			
KIND OF MATERIAL	**Conc. Form Lumber**	**Framing Lumber**	**Ready-Mi[x] Concrete**
PREVIOUS	2500 FBM	—	56 cy
TO-DAY	—	3010 FBM	9
TOTAL	2500	3010	65

PRACTICAL
FORM 110
MFD. IN U. S. A.

DAILY CONSTRUCTION REPORT

BUILDING Smith Store Building DATE June 6, 19--
LOCATION 656 N. Main St., Arlington, Ill. WEATHER Clear
ARCHITECTS Johnson and Anderson TEMPERATURE 62 7 A.M. 83 2 P.M.

CLASS OF WORK	Foremen	Mechanics	Laborers	Misc.
1. WRECKING—SHORING				
2. EXCAVATING—PUMPING		1	2	
3. PILING OR CAISSONS				
4. FOUNDATIONS—RETAINING WALLS				
5. WATER—DAMPPROOFING				
6. CEMENT FLOORS—WALKS—PAVEMENTS		2	2	
7. BRICK, TILE, CONCRETE MASONRY	1	4	6	
8. CAST STONE—CUT STONE—GRANITE				
9. TERRA COTTA				
10. FIRE-PROOFING, TILE—GYPSUM				
11. ARCHITECTURAL CONCRETE				
12. REINFORCED CONCRETE				
(a) Forms (b) Reinforcing Steel				
(c) Mix and Deposit Concrete				
13. ROUGH CARPENTRY	1	3	2	
14. FINISH CARPENTRY				
15. WOOD FLOORS				
16. INSULATION, SOUND DEADENING				
17. WEATHER STRIPS—CAULKING				
18. LATHING				
19. PLASTERING				
20. FIRE DOORS—WINDOWS				
21. STEEL DOORS—WINDOWS				
22. SHEET METAL WORK				
23. ROOFING				
24. TILE AND MOSAIC				
25. ASPHALT, CORK, LINOLEUM FLOORS				
26. ART MARBLE—SCAGLIOLA				
27. MARBLE—SLATE Exterior Interior				
28. GLASS AND GLAZING				
29. PAINTING AND DECORATING				
30. STRUCTURAL IRON—STEEL Erection Riveting		2		
31. MISC. IRON AND STEEL				
32. ORNAMENTAL IRON—STEEL				
33. PLUMBING—SEWERAGE—GAS FITTING		2	2	
34. HEATING AND VENTILATING				
35. AIR CONDITIONING				
36. ELECTRIC WIRING				
37. LIGHTING FIXTURES				
38. ELEVATORS—ESCALATORS				
39. SPRINKLER SYSTEM				
40. MAIL CHUTE				

REMARKS

2. Bulldozer operator and 2 laborers backfilling around foundation walls.

6. 2 cement masons and 2 laborers placing cement floor in basement.

7. Laying face brick and common brick back-up on east elevation. Laying solid common brick south wall.

13. Framing and placing joists for first floor.

30. 2 ironworkers finished erection of lally column and girders.

31. 2 plumbers working on connection to water main. 2 laborers backfilling sewer trenches.

Owner visited job today. Seems well satisfied with progress.

Must have 1"x6" D&M for rough flooring immediately. Lumber yard says they are tracing a car of this material which was due in two days ago. Better check with them.

Also check with heating contractor. Should have started today.

FRANK R. WALKER CO., PUBLISHERS, CHICAGO

FIGURE 7-4

Daily Material Report—Daily Construction Report

(Frank R. Walker Co., Publishers
5030 North Harlem Ave., Chicago, Ill. 60656)

DAILY JOB REPORT

DATE _____

CUSTOMER _____ JOB NO. _____

LOCATION _____ SHEET NO. _____

EMPLOYEES ON JOB	*CLASS	TYPE OF WORK	TIME	OFFICE USE

MATERIALS PURCHASES AND DELIVERIES

SUPPLIER	INVOICE NO.	✓ CASH	CH'G.	AMOUNT	

AUTO AND TRUCK EXPENSE _____
 (CASH RECEIPTS REQUIRED)

MILEAGE _____

SUBCONTRACTORS ON JOB _____

ESTIMATED JOB PROGRESS AND WORK REMAINING _____

*F FOREMAN   J JOURNEYMAN   H HELPER   O OTHER   TT TRAVEL TIME

PS 72007

FIGURE 7-5

Daily Job Report

Labor Report

EMPLOYEE	JOB NO. DESCRIPTION OF WORK	JOB NO. REMARKS	JOB NO.	WEEK ENDING					
				MON.	TUE.	WED.	THU.	FRI.	SAT.

FIGURE 7-6
Labor Report

DEVELOPING TIGHT JOB-SITE PROJECT CONTROL

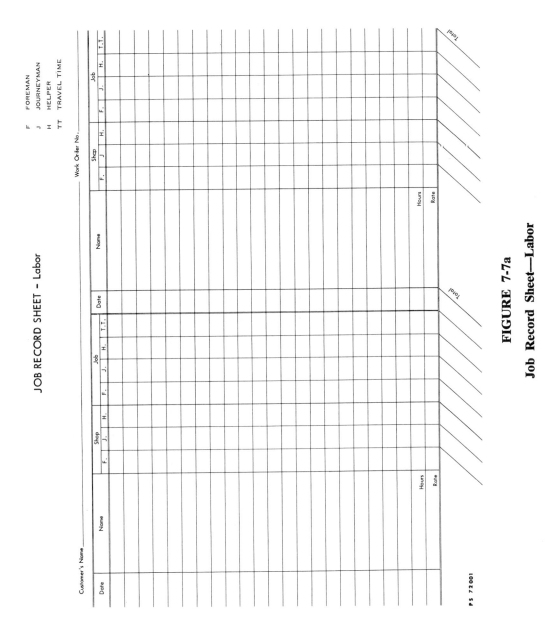

FIGURE 7-7a
Job Record Sheet—Labor

JOB RECORD SHEET

Sheet No. _____
Customer's Name _____ Address _____ Date Started _____ Date Tabulated this Sheet _____
Job Address _____ Work Order No. _____ Quoted Price _____ Tax _____
Charged to Date _____ Tax _____
Balance _____ Bal. _____ Total _____

Tabulation of Labor and Material	Remarks	Quantity	Description	Unit Price	Amount
Previous Labor Charged (Net)					
Labor This Sheet (Net)					
Total Labor to Date					
Previous Material Charged (Net)					
Material This Sheet					
Total Material to Date					
Total Labor Allowed					
Total Labor to Date					
(Net) Bal.					
Total Material Allowed					
Total Material to Date					
(Net) Bal.					

RECAP:

		SUMMARY	
Material (Net)		Material (Sell)	
Labor (Net)		Labor (Sell)	
Ins. %		Misc. Exp. (Sell)	
Misc. Exp. (Net)		Total	
Sub Total		Tax	
Overhead %		Balance	
Sub Total		Profit ÷ S.P. = ____ % Profit on S.P.	
Profit %			
Tax %			
TOTAL			TOTAL

FIGURE 7-7b
Job Record Sheet

DEVELOPING TIGHT JOB-SITE PROJECT CONTROL 105

Employee Time and Job Forms

Shown in Figure 7-8 is a Time and Job Ticket which has been designed and produced by the Reynolds and Reynolds Company, 800 Germantown, Dayton, Ohio

TIME AND JOB TICKET

DATE	JOB NO. OR DESCRIPTION	EMPLOYEE NAME OR NO.	HOURS WORKED	RATE	EARNINGS
			R		1
			O		
			R		2
			O		
			R		3
			O		
			R		4
			O		
			R		5
			O		
			R		6
			O		
			R		7
			O		
			R		8
			O		
			R		9
			O		
			R		10
			O		
		TOTAL REGULAR HOURS			
		TOTAL OVERTIME HOURS			
		TOTAL EARNINGS		$	

FOREMANS SIGNATURE

FORM T-1000

FIGURE 7-8

Time and Job Ticket

(Reynolds & Reynolds Co., 800 Germantown, Dayton, Ohio 45401)

45401. This is a three-part form interleaved with carbons. The third copy is a lightweight card stock which provides a firm support for making entries. The top copy is perforated along each horizontal line so that it may be separated into time on respective jobs and to accumulate time for up to ten employees. Spaces are provided on this ticket for as many as ten employees working under one foreman on one or more jobs. It is flexible enough to be used in several ways.

Another ticket, shown in Figure 7-9, was designed by the Reynolds and Reynolds Company for Binele Construction Corporation to be used in conjunction with their NCR bookkeeping machine. This form utilizes a two-part bottom stub uniset form with copy #1 perforated into four sections with each section consecutively numbered. The foreman on each job posts each man's time daily on the form by type of work performed. If a man works on more than one job during the same day or during the same week, the second perforated section would be used for the second job, the third perforated section for the third job, etc. The total time, rate and wages are recorded in the upper section by day. This section is used as source material for writing the payroll. The payroll department uses the top right section of copy #1 to record computed payroll data to be entered into the payroll records. At the end of the week, the employee checks his record and signs the form at the upper left to eliminate disputes in regard to the time worked and kind of work done.

When copy #1 is received in the office, it is torn apart with the upper portion of copy #1 used by the payroll department as indicated above. The three lower sections are also torn apart and sorted by job name for posting to the job cost ledgers. Part #1 is retained by the foreman to prepare his records and job progress reports. This ticket provides the control desired and reduces the time spent in posting to the job cost records. The payroll routine is also speeded and simplified.

Pocket Size Data Systems

Problem Solution Associates (see page 191) have what is known as the "Pocket System." This is a pocket-size ring binder in which any assortment of forms and paper may be inserted for field use. The ring binder is equipped with a ten-section index which can be filled out to suit individual needs. Numerous types of forms and other punched sheets are available, some of which are shown in Figures 7-10 (a, b, and c) and Figures 7-11a and 7-11b. You will note that legal forms such as waiver of lien, installment notes, and others are available. I can not overemphasize the importance of having the correct forms on the job at all times. Time cards, daily construction reports, purchase orders, and limitless others can be carried by the superintendent right in his pocket.

The Frank R. Walker Company (Chapter 12), which specializes in contractor forms, also has available many forms which fit the Pocket System Binder. If you are unable to locate the exact form for your purposes, send in a sketch to Problem Solution Associates, who will design the form without charge.

Another pocket-size system which includes a selection of several thousand different forms is Lefax. Their system not only offers numerous different forms, but also

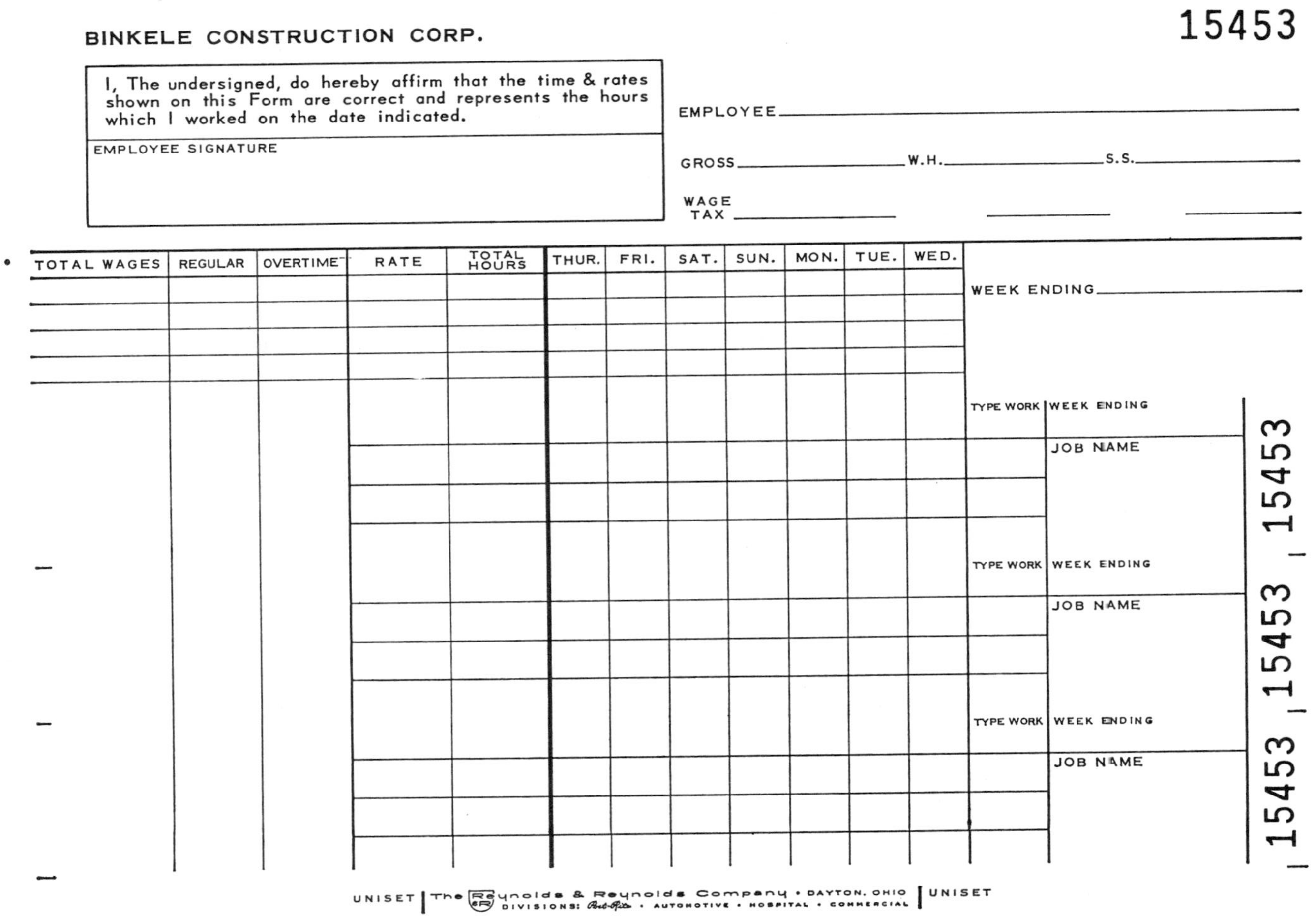

BINKELE CONSTRUCTION CORP. 15453

I, The undersigned, do hereby affirm that the time & rates shown on this Form are correct and represents the hours which I worked on the date indicated.

EMPLOYEE SIGNATURE

EMPLOYEE ____

GROSS ____ W.H. ____ S.S. ____

WAGE TAX ____

TOTAL WAGES	REGULAR	OVERTIME	RATE	TOTAL HOURS	THUR.	FRI.	SAT.	SUN.	MON.	TUE.	WED.

WEEK ENDING ____

TYPE WORK | WEEK ENDING | JOB NAME — 15453

TYPE WORK | WEEK ENDING | JOB NAME — 15453

TYPE WORK | WEEK ENDING | JOB NAME — 15453

UNISET | The Reynolds & Reynolds Company • DAYTON, OHIO | UNISET
DIVISIONS: Post-Rite • AUTOMOTIVE • HOSPITAL • COMMERCIAL

FIGURE 7-9

Employee Time and Job Form

(Reynolds & Reynolds Co., 800 Germantown, Dayton, Ohio 45401)

PROPOSAL

_____ 19____

TO_____

_____

_____ propose to furnish all materials and perform all labor necessary to complete the following:

_____
_____
_____
_____
_____
_____
_____
_____
_____
_____

All of the above work to be completed in a substantial and workmanlike manner for the sum of

_____ Dollars ($_____)

Payments to be made each _____ as the work progresses to the value of _____ per cent (_____%) of all work completed. The entire amount of contract to be paid within _____ days after completion.

Any alteration or deviation from the above specifications involving extra cost of material or labor will only be executed upon written orders for same, and will become an extra charge over the sum mentioned in this contract. All agreements must be made in writing.

Respectfully submitted,

_____

ACCEPTANCE

You are hereby authorized to furnish all materials and labor required to complete the work mentioned in the above proposal, for which _____ agree to pay the amount mentioned in said proposal, and according to the terms thereof.

_____

P 2006

FIGURE 7-10a

Proposal and Acceptance

DEVELOPING TIGHT JOB-SITE PROJECT CONTROL

PURCHASE ORDER

No. _____ Date _____ 19 _____

Job _____ Job No. _____

To _____

Address _____

PLEASE DELIVER THE FOLLOWING ORDER TO:

QUAN.   DESCRIPTION   PRICE

INVOICES MUST STATE ORDER NUMBER AND PLACE DELIVERED.
PRICES ON THIS ORDER NOT SUBJECT TO CHANGE.

FORM P-113   PURCHASER

FIGURE 7-10b

Purchase Order

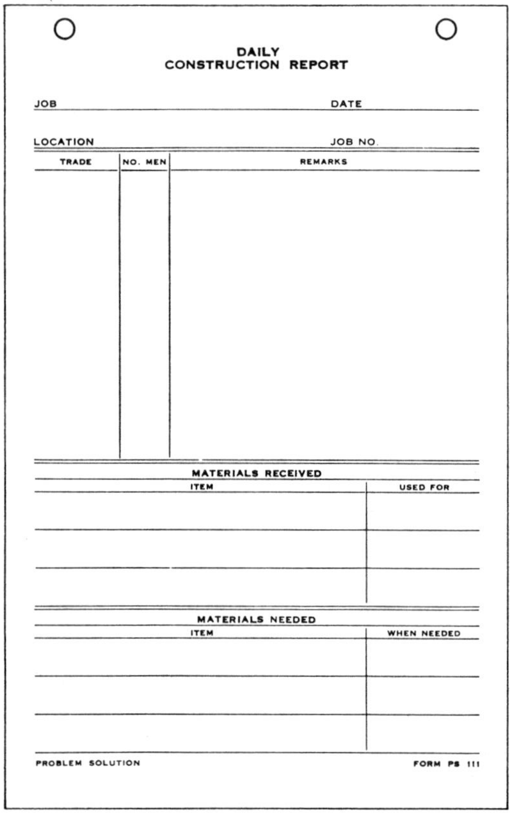

DAILY CONSTRUCTION REPORT

JOB DATE

LOCATION JOB NO.

TRADE	NO. MEN	REMARKS

MATERIALS RECEIVED

ITEM	USED FOR

MATERIALS NEEDED

ITEM	WHEN NEEDED

PROBLEM SOLUTION FORM PS 111

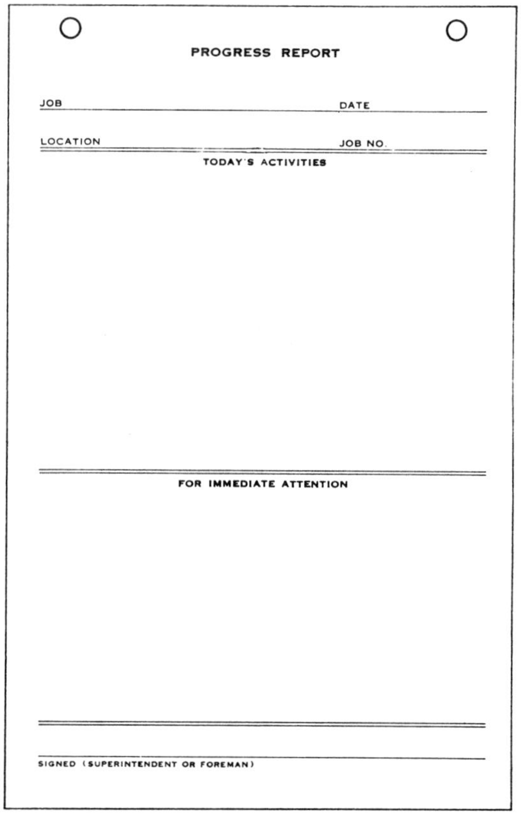

PROGRESS REPORT

JOB DATE

LOCATION JOB NO.

TODAY'S ACTIVITIES

FOR IMMEDIATE ATTENTION

SIGNED (SUPERINTENDENT OR FOREMAN)

FIGURE 7-10c

Daily Construction Report and Progress Report

makes available engineering information for almost every branch of engineering. Some of the available forms are shown in Figures 7-12a, 7-12b and 7-12c.

You will note that the Lefax sheets are punched for side rings and that top rings are used in the Problem Solution Pocket System and the Frank R. Walker system; therefore, the sheets are not interchangeable. Consequently, you must select one style and stick with it. Sheets for all three systems are $6\frac{3}{4}''$ long; both the Problem Solution and the Walker systems are $4\frac{1}{4}''$ wide, whereas the Lefax system is $3\frac{3}{4}''$ in width. Ask for catalogs and sample sheets from the source of your choice.

DEVELOPING TIGHT JOB-SITE PROJECT CONTROL 111

FIGURE 7-11a

Installment Note

FIGURE 7-11b

Waiver of Lien

CONSTRUCTION REPORT

Job..

Building..

Date

LEFAX FILING INDEX

Excavation................................Yds. Completed..................

Footings Poured for Col. Nos...

....

Columns No. Poured or Erected..

....

Beams Poured or Erected...

....

Sq. Ft. of Slabs Poured........................Floor No...................

Sq. Ft. of Flooring Laid.........................Floor No...................

Forms Complete..

Forms Stripped..

Will Pour...Tomorrow

Sqs. of Roofing Laid.....................Complete..............................

Brickwork...

....

Sash Erected..

Doors...

Material Received	Amount	Inspected

In order to speed up work we need..

....

....

....

Remarks..

....

....

....

By..(Engineer)

Insert temperature and either Snow, Rain or Fair

Weather	A. M.	P. M.

FIGURE 7-12a

Construction Report

(Copyright by Lefax Publishing Co.,
1315 Cherry St., Philadelphia, Pa. 19107)

Total concrete poured...cu.yds
Cement finished..sq.yds.
Sacks of cement used..
Lime used...
Number of bricks laid.............@.............or.............per M

Men	Trade	Hours	Wages	Remarks
....	Supt.			
....	Foreman			
....	Teamsters			
....	Bricklayers			
....	Carpenters			
....	Plumbers			
....	Plasterers			
....	Painters			
....	Roofers			
....	Laborers			
	Totals			

Clearing Site............	....	Woodwork............	....
Excavating–Grading....	....	Roof Covering........	....
Reinforcing...............	....	Lath–Grounds........	....
Concrete...................	....	Plaster...................	....
Forms.......................	....	Plumbing................	....
Structural steel.........	....	Steam.....................	....
Brickwork.................	....	Electrical installation	....
Stone–Terra Cotta....	....	Marble...................	....
Cutting Brick–Stone	....	Glazing...................	....
Sifting sand..............	....		
Scaffolds..................	....		....
Framing....................	....	Total	....

Special Instructions from Architect's Representative

....
....
....
....
....

FIGURE 7-12a (cont.)

Daily Construction Report

TR.MK. REG. U.S. PAT. OFF

Location..Date............

....Weather............

Equipment	Hrs.	Amt.	Labor	Hrs.	Amt.
Total			Total		

Material Delivered	Quantity	Material Delivered	Quantity

Work Done:..

LEFAX, PHILA. 19107
MADE IN U.S.A.

L-1369

FIGURE 7-12b

Daily Construction Report

(Copyright by Lefax Publishing Co., 1315 Cherry St., Philadelphia, Pa. 19107)

	DAILY REPORT

Date:
Job No:
County:
State:

TR.MK. REG. U.S. PAT. OFF.

Location _____

Name _____

Description of Work Being Done

Name	Occupation	Hrs.

Weather Report

A. M.

P. M.

FIGURE 7-12c

Daily Report

(Copyright by Lefax Publishing Co.,
1315 Cherry St., Philadelphia, Pa. 19107)

8

Purchasing, Expediting, and Preventing Loss of Materials

New materials are being introduced almost constantly and research projects are under way which will result in expanded uses for present materials and lower costs of building and construction. It will certainly pay to keep abreast of these important developments. Building codes which have severely restricted the use of new materials and methods are being relaxed and re-written to permit new products and improved construction systems to be utilized. Pressure has been constantly applied by associations, manufacturers, contractor groups, and the government agencies associated with the building and construction industries. One significant example is the use of wood in construction.

FIRE-RETARDANT WOOD

The Uniform Code has now approved fire-retardant woods for exterior walls and the Southern Code has approved fire-retardant woods for exterior non-bearing curtain walls in high rise buildings of any type of construction in all areas. Also, FHA has now allowed the use of fire-retardant lumber for structural members in multifamily housing and many commercial buildings where steel was formerly required. This substitution can amount to considerable savings in construction costs.

Fire-retarding chemicals, usually consisting of ammonium salts compounds, are impregnated under high pressure into wood fibers. The fire-protection chemicals react at temperatures below the ignition point of untreated wood and produce a nonflammable vapor, water, and carbon. This reaction reduces the flammable gases and tars and forms a protection char which insulates the surface of the wood. By protecting against further burning, the charred surface protects the strength of the wood beneath. The chemical treatment prevents the spread of flames and eliminates afterglow. One of the problems with material used in nonflammable structure is that it deforms and collapses under high temperatures, whereas properly treated fire-retarding lumber will actually withstand higher temperatures without collapse.

Extensive research has been conducted by the U.S. Forest Products Laboratory, Madison, Wisconsin. Western red cedar shingles have been fireproofed to the extent that they are expected to be effective for 35 years or more! One of the leading producers of fire-retardant and other pressure treated lumber is the Koppers Company.

Fire-retardant acoustical tile using wood fibers has been developed and additional savings in construction costs may be realized. This is another interesting product which may be considered for new construction and remodeling.

SAFEGUARDS IN THE PURCHASE OF MATERIALS AND COMPONENTS

To be assured of receiving materials of specified quality, it is best to request samples. These samples may or may not be put under laboratory tests, but the fact that suppliers have been requested to submit them has a tendency to keep them from making inferior substitutions. Keep the samples until the project guarantee has expired. If material different from the samples has been used on the job, it can usually be established by comparison with the samples. In cases of protective coatings, apply some of your sample on the project and make notes of the test area for future comparison.

It is often desirable to have laboratory tests made to confirm quality. This decision must be made in the light of surrounding circumstances. If you have any reason to question the quality, by all means have it tested. Make out a list of samples to be requested and the quantities necessary for the tests, and, where possible, request the samples before contracts or purchase orders are awarded.

STORAGE AND PROTECTION OF MATERIALS

Since the cost of materials represents a very substantial percentage of the project cost, every consideration should be given to the safe handling and storage of every item brought on the job. It is necessary to provide protection from the elements, but it is equally important that consideration be given to minimizing damage during construction activities. Consideration must be given to the location of materials to provide the best protection from vandals and pilferers. A construction project provides a natural attraction to children and much damage can be done if they are given access to materials.

If materials are stored within the building, the weight should be estimated and the load spread so that the structure is not overloaded. This is one of the requirements of the Occupational Health and Safety Act. If materials are placed on or near public thoroughfares, red lights must be kept burning at night and red flags posted during the daylight hours.

Don't forget that cement, lime, and plaster should not be placed in the open, especially if humidity exceeds 50 percent. Humidity usually increases considerably at night. Even if the material is covered and it does not rain, slight absorption of moisture can begin chemical reaction resulting in deterioration. Even a limited amount of moisture absorption by cement will affect the yield strength of concrete. Also, the partial absorption of moisture by lime causes partial carbonation resulting in reduced strength of the mortar or plaster in which it is used.

Problems can develop if aggregates are left exposed to the weather. If weather has been below freezing, ice formation on the surface of aggregates will affect the adhesion of cement. Also, the amount of mixing water must be adjusted if sand and gravel become wet.

Some lightweight aggregates will become crushed at the bottom of the pile if piled more than four to five feet high, and crushing will reduce the effectiveness of these aggregates.

Don't forget to protect sand and lightweight materials from high winds. This will not only conserve materials, but will also provide safety to personnel and equipment.

Density of concrete or masonry block and brick varies considerably. Some of the more porous materials will absorb more moisture than desired if left exposed to the elements, and if they are placed on wet soil, algae will form which will discolor them and ruin their appearance. Algae *can* be removed, but it is time consuming. The method for removing algae is described in a report prepared by Problem Solution Associates (see page 179).

Although it may seem unnecessary to caution against damage to masonry components, much damage is done to these expensive materials on almost every project. Quite often, damaged brick or block will be used by turning them so that the chipped sections or cracks are not visible. This construction may result in leaking walls and affect the strength of the structure.

A moderate amount of rust on steel re-bars will improve the adhesion and loosen scale, so it it not necessary to provide protection against the elements for the duration of the average construction job. Slight corrosion of structural steel members is not particularly harmful. Just be careful not to pile steel in areas containing salt water, chemicals, ashes or other corrosives.

Materials cannot always be stored inside for protection and must be covered in some manner outside. One of the most satisfactory protective covers is a product known as "VersaTarp," which is a plastic tarpaulin reinforced with fiberglass mesh. It is tough and will withstand high winds. Information is available from the Griffolyn Company, Inc., 10020 Mykawa Rd., Houston, Tex. 77048.

HOW TO REDUCE THE LOSS OF MATERIALS

There are two basic types of losses of materials which can be very costly. First is loss through theft and pilferage. Protection of materials from these unnecessary losses is essentially the same as for tools and equipment. This subject is treated thoroughly under the heading "Reducing Loss of Tools, Equipment and Materials," in Chapter 9.

Next is the loss of materials through carelessness, improper handling and storage, or improper use. These problems can only be corrected by education of foremen and management personnel. As with most such problems, the problem starts at the top. Supervisory personnel must be trained in the conservation of materials.

The best approach is to make a survey of the wastes on a specific project, writing down each instance which can be observed. Right here is a good application of the camera. Time lapse photos can also be effective in pinpointing waste of materials. (See "How Costs Can Be Lowered by Methods Engineering," Chapter 5.) Take a photo of the situation that needs correction. After all the evidence has been collected, set up a meeting of the foremen and other supervisory personnel concerned. Go over the items of waste, one by one. Do not condemn; instead, ask the group to suggest possible reasons for the problem and for ways to correct it. By taking this approach, many of the actual reasons and causes will be brought to light. For example, new interior grade plywood was used to cover a stack of cement. The reason may have been that plastic material normally used could not be located. If plenty of plastic was on the job, what happened to it? It may develop that someone had moved it and failed to return it to its proper place. By this line of exploration, many small discrepancies in procedures may be revealed, leading to a considerable number of unexpected gains.

LUMBER AND MILLWORK

Quality and characteristics of these materials may vary more than in any other construction material. Purchase of lumber and millwork from small local mills is risky and more investigation is essential than in making purchases from large established sources. Quality of materials can be determined by visual inspection, but moisture content must be checked with a moisture meter. I have frequently carried a meter with me so that I could check the moisture content. In most instances, moisture was nowhere near what it was claimed to be!

Presented in Figure 8-1 is a geographical chart of the United States showing recommended average moisture content for interior finish woodwork in various parts of the country. You will note that this varies from 6 to 11 percent in different sections of the country. The moisture content of woodwork to be installed should be within about 5 percent of that recommended in order to avoid the risk of joints opening up and possible checking of wood. This ideal condition is seldom met, as can be witnessed a year or so after completion of most jobs. In a mild, relatively dry climate, the problem is not so prevalent.

The purchase of lumber is a very important item. If lumber is below grade, excessively warped or wet, the cost of installing it can jump 20 percent above estimate; much worse cases have been seen frequently. Consequently, it will be a good investment in time to make an investigation of the quality before the material is delivered, if possible. By all means, stack it properly, on a level surface and out of the weather. Mills with membership in one of the lumber associations usually mark the grade on the material.

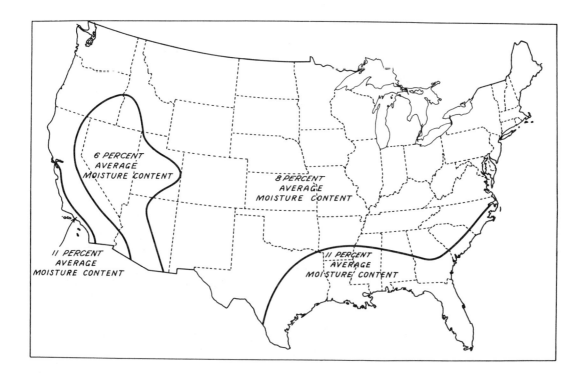

FIGURE 8-1

**Recommended Average Moisture Content
for Interior Finish Woodwork
in Various Parts of the United States**

It is suggested that this grading code and specifications of the various grades be obtained from the association. For your convenience, a list of the various asociations and their addresses are included in the Appendix.

In purchasing cabinetwork, it is best to visit the shop or mill to check into construction used. Facings, for example, should be tenoned or doweled at the intersections. This is a rather costly operation and quite often is omitted. Try to twist the facing to see if it moves at the joint. If it does, the joints have not been properly made.

Framing lumber is normally about 15 percent moisture content, but this figure can vary considerably and should be checked with a moisture meter. *Even at 15 percent moisture content, lumber has only shrunk about half its maximum!* From this it is obvious why shrinkage cracks occur later. Millwork lumber should be 6 to 10 percent moisture content, depending upon the geographical location.

All wood products, even if primed, should be protected against the weather. They should not be allowed to absorb too much moisture and they should not be subjected to too much drying from the sun or other source of heat. Because of the possibility of staining, be sure that creosoted or treated lumber is not stored adjacent to untreated lumber or other materials.

Wood flooring installation must be handled very carefully to prevent cracks after occupancy. The drying processes of concrete, plaster, and even water-base paint raise the humidity in a building to the point where flooring materials stored in the building will absorb excessive moisture. Even prefinished flooring and paneling absorb moisture under such conditions, but at a slower rate. By all means, do not accept materials which have become wet during delivery. If materials with a high moisture content are installed, shrinking and open joints are inevitable.

During drying periods of the building, it is important to provide ventilation and, if possible, heat. Interior woodwork should not be brought on the job, or at least not into the building, until ready for use. Moisture content must be within the proper range, depending upon the section of the country where located (see Figure 8-1). The only satisfactory method of determining moisture content is with a moisture meter, which would certainly be a good investment for a building contractor.

It is recommended that trim materials and flooring be spread out for at least four days to expose all surfaces to the air, providing that the building is heated to reduce humidity. Temperature within the building should be kept at near 70°—and somewhat higher will do no harm. Another advantage in keeping interior woodwork dry is that sanding and painting are easier and a better job results.

The use of particle board (pressed wood chips) for long shelves should not be permitted, or if the shelves are to be heavily loaded, particle board is not recommended. Personally, I would not use it for shelving lengths more than 36" long. The material is difficult to finish properly, so do not allow its use for exposed surfaces. A plastic edge finishing strip is available for improving the appearance of the shelf edges.

PLACEMENT OF MATERIALS AT THE JOB SITE

Considerable thought must be given to the location of materials and the piling of excavated earth so that they do not have to be moved later. Before placing any material, check to see if the area is to be available for the duration of the job. Will the area have to be reached for sewers, septic tank, water, or other utility service? Will television conduits or telephone lines have to go in? Will it have to be used for a drain field? Will it be in the way of scaffolding, cranes, driveways, walks, or construction activities? Is the contemplated location as near to the point of use as possible?

I have witnessed many costly problems in the making. Once, as I was discussing a technical point with a construction superintendent, one of the men interrupted by asking: "There is a load of brick here. Where do you want it unloaded?" The superintendent said, "Oh, just anywhere out of the way."

PURCHASING, EXPEDITING, AND PREVENTING LOSS OF MATERIALS 123

After we had completed our discussion and walked outside, the superintendent was amazed to see that the brick had been placed so far "out of the way" it would have to be relocated. Also, grading of the area was planned for the following morning, and equipment was not available for handling it.

Have you considered the idea of taking a plat map and showing all utilities and services to be installed and indicating the space available for locating materials? Some preliminary thought on this subject could save valuable time and much inconvenience later.

How to Avoid Job Delays Due to Materials and Supplies Problems

Quite often, materials arrive on the job site, are unloaded, and errors in shipment are not discovered until time to use them. This carelessness can wreck a schedule many times on the same job. To avoid this problem, use the checklist below and carefully check all deliveries for quantity, condition, and accuracy. When possible, check deliveries before unloading. In cases where this is not feasible, make the investigation immediately after unloading.

Materials Checklist

1. Is this the correct type of material as specified?
2. Is this the exact color required?
3. Check all dimensions for correct size.
4. Inspect all visible areas for damage.
5. Is the order complete? Any items back-ordered should be placed on the ALERT list for immediate follow-up.
6. Absorptive materials such as wood or cement should be checked to see that they have not become wet.
7. Are the materials being placed as close to the point of use as possible?
8. Will the materials be safe from the weather, damage, theft, and other hazards?
9. Are they being placed so that they can be easily handled and moved if necessary?

UNAUTHORIZED SUBSTITUTION OF MATERIALS

A few contractors and subcontractors attempt to substitute the cheapest materials they can possibly locate. When a contractor knowingly bids the job too low he hopes he can cheat his way along, one way or another, and wind up with a profit. Usually this type of operator doesn't last long, and he seldom profits from cheating, but he does considerable damage while he is active. It is well to be aware of these tactics and guard against them.

A certain amount of cheating has been discovered in almost every segment of the industry—to the point where architects and engineers have learned to distrust just about everyone. One contractor I knew would obtain a five-gallon can of the material specified and then refill the can from other cans of a much inferior material. He probably lost more time running out to his truck or storage area to refill that one can than he saved, but he seemed to get a kick out of cheating. We solved that problem by taking a small sample from each can full of material regardless of the label on the can. It wasn't actually necessary to test the material. When we came for the sample, he admitted the whole scheme and gave up the cheating—at least on the subject project.

An architect specified tempered glass on a project and asked us how he could determine whether the glass was actually tempered after installation. We advised striking the glass with a rubber mallet (see our method of testing for tempered glass in Chapter 11). It turned out that about half of the glass was tempered and the other half shattered under the blow of the mallet, indicating that it was not tempered. The subcontractor in this instance had to replace half of the glass at his own expense with tempered glass. Let's hope that this expensive lesson has cured him of cheating.

We are all familiar with the cheating in concrete and plaster mixes. This risk can be greatly reduced by having a preliminary meeting with the subcontractors in which problems and their causes are discused. If they know that project managers and inspectors are going to be watching for symptoms of deviations from specifications, they will be more likely to "toe the mark." Most problems of this type can be prevented by diplomatic discussions or meetings with all suppliers and subcontractors at the beginning of the project. Careful selection in the first place can eliminate most of the cheaters.

PROPER ALLOWANCE FOR LOSS WHEN FIGURING MATERIALS

The practice of figuring materials too closely is usually false economy. I could cite numerous instances of the costly repercussions due to this practice. On one job using an expensive tile, the contractor ordered only a few square feet more than the actual area covered. After the tile setters broke a few pieces, they ran short. The replacement order did not match, and a second replacement order did not match. The tile contractor finally had to replace the entire installation, delaying completion of the project. If he gained anything on previous jobs by figuring close, he certainly lost it all on this one!

The lesson to be learned from this is that the risk and consequences must be considered in each instance. Colors of different batches of materials usually vary. If more material from the same batch is in stock, there is little risk in figuring closely. On the other hand, if material must be ordered from someone else's stock, you run the risk of depletion of the batch your order came from by the time you need extra material. A careful analysis of past experience is important here. If you do not have your own experience as a guide, do a little research and benefit from the hard-earned experience of others. Hasty guessing can be very costly.

PROTECTIVE COATINGS FOR STEEL

Many of the more chemical-resistant and water-resistant coatings now available possess poor surface penetrating and bonding characteristics, and therefore require careful preparation of the surfaces to be protected. Since formulation of the coating will be greatly influenced by the color, no attempt will be made here to give specifications. However, a substantial percentage of basic lead silico chromate is recommended, particularly in the primer and undercoat.

In preparing the surfaces, it is not necessary to remove all paint, scale, and rust, but all loose material and foreign matter must be removed to provide a solid foundation for the new coating. Here is a brief guide to good preparation of steel surfaces for painting:

1. Remove all oil and grease with solvents and rags.
2. Scrape off all foreign matter down to the bare metal.
3. Remove all loose scale by hand scraping or power equipment, such as sand blast, needle scaler, wire brush, or other equipment.
4. Heavy edges of old paint should be removed by one of the methods mentioned in item 3 above, and feather-edged with sand paper.
5. A final cleaning should be done with a non-oily solvent such as naphtha, and clean rags.
6. Be certain that all surfaces are dry before applying primer.

More information is available from National Lead Industries, Inc., 111 Broadway, New York, N.Y. 10006. Detailed specifications and recommendations for preparing steel surfaces and application of protective coatings are available from Steel Structures Painting Council (see Appendix for address).

9

How to Get More from Your Tools and Equipment

It is generally assumed in the construction industries that tools and equipment do not last long. This is a costly assumption because it psychologically conditions personnel to be careless and inconsiderate in their handling, operation and care of equipment. It is granted that conditions of use in all types of weather—dirt, sand, rain, and other corrosive elements—can markedly shorten equipment life; however, in most instances, the useful life of equipment could be greatly extended by inauguration of a well-planned equipment program.

FIVE-STEP PROCEDURE FOR A WELL-PLANNED EQUIPMENT PROGRAM

What do we mean by "equipment program"? Here is an outline of the essentials of such a program:

• 1. *Set up an equipment file for every piece of equipment.* The file should contain everything that may be needed on the history of the item, including cost of equipment and attachments, suppliers and individuals you deal with, operating instructions, parts lists, complete history of repairs, supplier's literature describing the equipment, and perhaps your own photo of the equipment. Other data such as insurance, licenses, permits, and compliance with safety laws, should either be included in the file or referenced as to location of such data.

- 2. *Work out an educational program to teach employees the proper use and operation of the various types of equipment used by your company.* This will include everything from office machines to heavy equipment. It is especially important to have a thorough program of proper use of equipment which could be dangerous to life and property.

- 3. *Establish a preventive maintenance program and schedule every piece of equipment for periodic inspection and repair.* This greatly reduces the chance of equipment failure at a critical time. All equipment should be included, even office machines. Assign this responsibility to the proper individual and follow through to see that the program is being followed as planned. It is a well-established fact that it is false economy to run equipment as long as it will go. The breakdowns usually occur at the most inopportune times, causing work stoppages, frustration, higher repair costs, and other intangible losses. Usually this type of breakdown causes unnecessary wear on other parts which might not be affected if repairs were made earlier. Proper periodic inspection usually reveals sources of potential trouble. Routine inspections, for example, may reveal a gasoline leak which could waste gasoline and possibly start a fire. Or a hydraulic leak may be detected which could later cause brake failure resulting in a serious accident.

- 4. *Set up depreciation schedules and a replacement program.* An essential part of this phase of the program is a regular, systematic investigation of the condition of the equipment to determine if it should be replaced, overhauled or eliminated. Too much down time or too slow operation may indicate that the equipment is not efficient and some type of action is needed.

- 5. *Devise an equipment scheduling system.* Obviously, equipment is of little value if it is not at the right place at the right time. And quite naturally, it can't be used in two places at the same time. Consequently, efficient equipment utilization depends upon proper planning for its future use. An "Equipment Schedule Board" may be laid out to suit your own purposes. On this board would be posted the location of each piece of equipment, planned use, and scheduled servicing and overhaul. We know that all equipment must be serviced, so why wait until it breaks down out on some remote job? If preventive maintenance is not properly scheduled, on the job scheduling will be of little use, since the equipment may not be available when planned due to down time for repairs. In any system, someone must be assigned the responsibility for the planning and scheduling and for posting this information constantly up to date.

Equipment Records

A complete system of forms for recording equipment operating costs and other important information is included in this chapter. First is the "Equipment Cost Daily Record" form shown in Figure 9-1. This form fits the Pocket System notebook described on page 106. It is used by the equipment operator to report all operating and repair costs. These daily reports are then transferred to the monthly "Time and Cost Record" shown in Figure 9-2. Totals are brought down at the end of each month and the summary finally

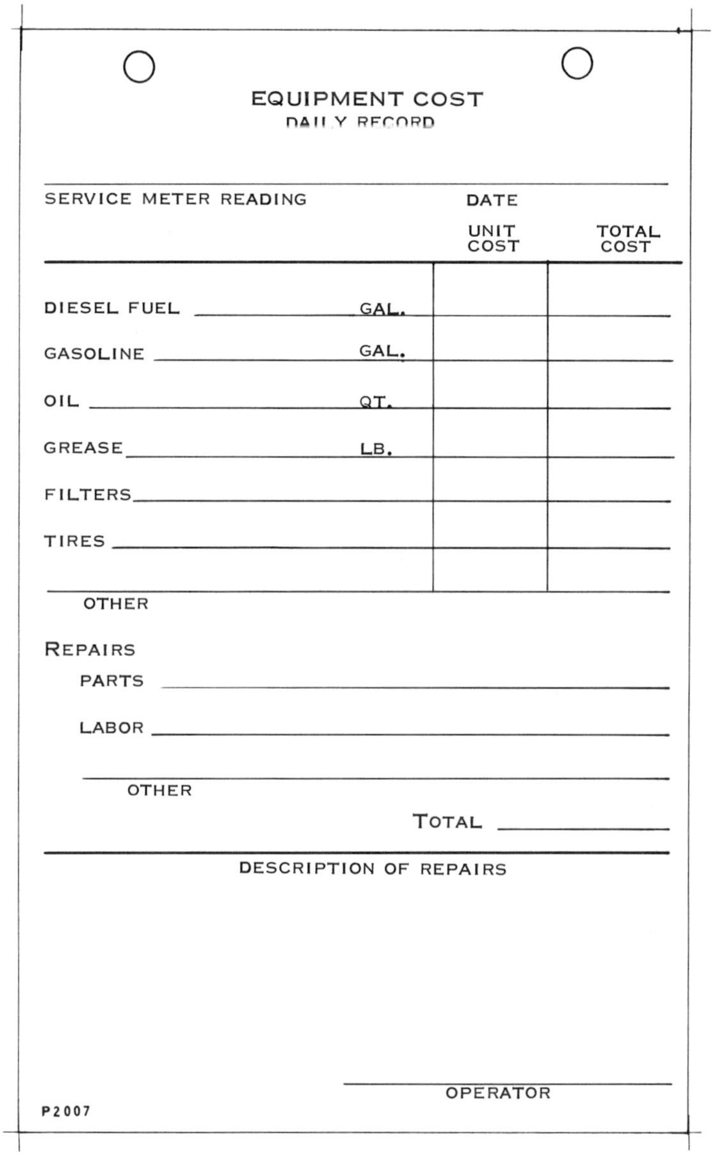
EQUIPMENT COST
DAILY RECORD

SERVICE METER READING DATE

	UNIT COST	TOTAL COST
DIESEL FUEL ______ GAL.		
GASOLINE ______ GAL.		
OIL ______ QT.		
GREASE ______ LB.		
FILTERS ______		
TIRES ______		
______ OTHER		

REPAIRS
PARTS ______
LABOR ______
______ OTHER

TOTAL ______

DESCRIPTION OF REPAIRS

______ OPERATOR

P2007

FIGURE 9-1

Equipment Cost Daily Record

inserted on the "Annual Summary Sheet" shown superimposed on the "Time and Cost Record" form.

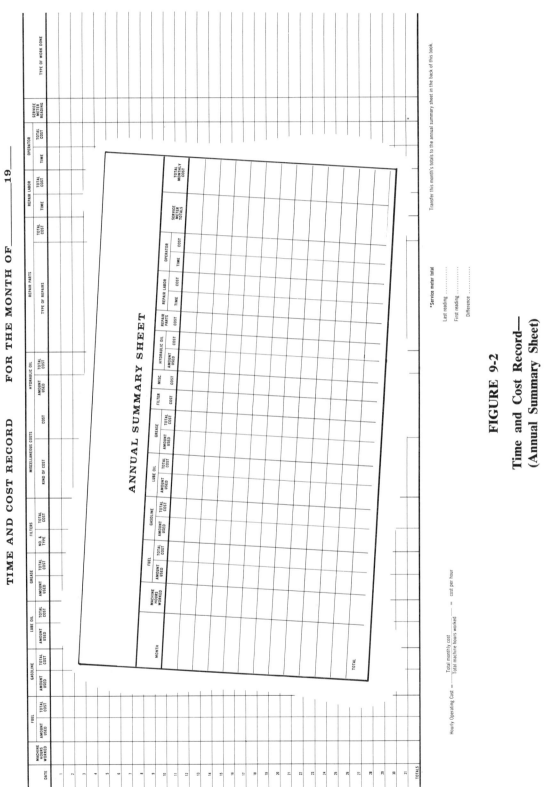

FIGURE 9-2

**Time and Cost Record—
(Annual Summary Sheet)**

(Caterpillar Tractor Form)

Equipment Charges to Projects

The purpose of the record form shown in Figure 9-3 is to provide a means of keeping up with the time each piece of equipment is used on each project so that costs can be accurately distributed to the job cost records. Many contractors are not keeping accurate records of the amount of time each piece of equipment is used on each project; consequently, job cost information is not accurate. Since equipment cost is an important part of the total job cost, accuracy of reporting is essential. This system will provide the necessary accurate information.

Regardless of the system you devise, its chances of success are in proportion to the follow-up given by top management. When you put new procedures into operation, see that all concerned are thoroughly familiar with the objectives and that they understand what is expected of them; then, observe the progress regularly to see that the program is on course. If not, find out why not. Many people balk at new procedures and find every excuse for their failure to comply. Make changes only after your are thoroughly convinced that change is necessary.

Assign one individual the full responsibility for managing the equipment and tools. His job may be full time or only part time, depending upon how much equipment is involved. The following records must be kept if efficient tool and equipment management is to be attained:

1. Location of all equipment and tools at all times.
2. Operating condition of all equipment.
3. Availability of each piece of equipment and every tool.
4. Compliance with all safety laws.

HOW PROPER TRAINING CAN REDUCE EQUIPMENT COSTS AND IMPROVE EFFICIENCY

Regardless of the pressure of time, it is almost always more expedient to give operators the proper amount of training on new equipment. You can't afford to take their word for their knowledge and experience. And this applies to the very simplest equipment. One contractor we knew employed a girl as a typist and secretary. The employment agency had assured him the girl rated very high. He asked the girl if she was familiar with the IBM proportional spacing typewriter. Not wishing to convey any negative impression right at the start of her job, she said, "Oh, yes!" She was placed in a small office, given a pile of work, and promptly forgotten by the contractor who left for a trip to his jobs. When he came in at 4:30 P.M. the girl was so distraught she was crying. Although she had had many years of experience with office equipment, she had never even heard of a proportional spacing typewriter!

A few minutes of instruction would have saved that situation. When the same situation arises out on a job the costs could be considerably higher. I remember the time when a heavy equipment operator who claimed to be experienced in the operation of a certain piece of equipment actually didn't know one lever from another! After wrecking

EQUIPMENT CHARGES TO PROJECT NO. ______ NAME ______ ADDRESS ______ MONTH OF ______

☐ COMPANY OWNED ☐ RENTED ☐ LEASED PROJECT MANAGER ______

EQUIPMENT	SERIAL NUMBER	DAY 1	2	3	4	5	6	7	8	9	10	11	12	13	14	15	16	17	18	19	20	21	22	23	24	25	26	27	28	29	30	31	HOURS USED	RATE PER HR.	TOTAL CHARGES
TOTALS																																			

FS 72101

FIGURE 9-3

Equipment Charges to Project

a few gears he was fired for his gross misrepresentations. But who paid for the repairs and three days' down time? The contractor, of course. Don't make the mistake of thinking that a certificate from some operator training school automatically qualifies a man on your equipment. New operators should be checked out by someone you know to be capable of giving the instruction. A very capable operator may not be qualified to instruct.

It is always best to have all instructions in writing. Usually, operating instructions are available for all equipment; however, quite often, the instructions are not clearly written or properly illustrated. In this case, it may be necessary to provide supplementary sketches or explanation. You can easily make your own photos to clarify your instructions. We have found that the best schooling is usually that offered by the manufacturer of the equipment. Quite often, your dealer can arrange to have factory instructors in for a training course in your local area. Ask him.

PSYCHOLOGICAL FACTORS AFFECTING THE LIFE OF EQUIPMENT

One of the most important factors which determine the care and treatment employees give their equipment is their *attitude toward the company*. In other words, their morale. If the employee is unhappy or irritated with his job, his supervisor, his working conditions, or any one of many possible frustrations, he will reflect this in the care and handling of the equipment he uses. If he is doing maintenance or repair work, the results are even more serious, since he may neglect or sabotage all of the equipment with which he comes in contact.

This disastrous attitude may be more or less subconscious, but in any case, he is never able to put forth his best efforts while his morale is low. You are probably saying, "So what? What can we do about it?" The one thing that can be done to improve this general condition is to begin training all management people in the human relations aspect of their jobs. The construction industry is lagging behind all others in management development. Of course, there are reasons for this. Let's take a look at them:

1. The *production* of a building or construction project is always located away from the office, whereas a commercial or manufacturing operation has most if not all of the management people in a convenient group. It is comparatively simple to call a management meeting for the purpose of handling management training or specific problems. The construction company, on the other hand, has more difficulty getting the management personnel together. Obviously, much depends upon the size and scope of the firm's operations.
2. Each construction project is different and presents a strong challenge. Construction management personnel are men of action! The atmosphere of the average construction project does not lend itself to the pussy-footing techniques of a department store or a manufacturing operation. The construction job is a tough one. Construction men are experienced in fighting the elements, rough terrain, and a never-ending series of obstacles. It is only natural, then, that a foreman expects more from the men in his "gang" than a commercial or manufacturing supervisor would of the "personnel in his department." A comparison of the terminology used is indicative of the tougher

approach taken in construction work. The construction worker, for example, may refer to a well-dressed man in a suit as "that guy in the monkey suit." He feels that he is doing the work while others are merely bystanders. I know, because I came up all the way through the ranks. I started as a water boy at the age of eight. At age ten I was pulling nails and cleaning forms. By the time I was 18, I was in charge of construction. I have had an opportunity to view all aspects of the jobs, from all viewpoints.

Back in the thirties, employees worked under much more pressure than today, but life was not as complex either. The "bull of the woods" technique has been superseded by tact and diplomacy, and the sooner construction management learns and employs these techniques, the smoother operations will become.

These psychological factors which affect the life of equipment also affect productivity, labor turnover, absenteeism, safety, materials waste, etc. More information on this subject is contained in Chapter 2, Organization and Management of Human Resources.

IMPROVING EFFICIENCY OF EQUIPMENT

When evaluating older equipment for replacement or modernization it is well to obtain firsthand information from the operators. Determine the nature and extent of the operators' problems. Quite often, operators feel that the frustrations and inefficiencies they encounter are conditions they "just have to live with," and they may never register any complaints. Much depends upon the attitude of management about listening to operators' complaints concerning equipment. Operators know that if repairs or service are needed they are to write up these items; but, very likely, they will not complain about levers which are difficult to reach, uncomfortable seats, noise, and other annoyances. These are the factors which greatly limit productivity. Heavy equipment often requires a "heavy hand" to operate. By the end of the day the operator is running at about half of normal productivity.

Power steering kits are available for converting vehicular equipment. One source is Garrison Manufacturing Company, Inc., 4609 E. Shelia St., Los Angeles, Calif. 90022.

REDUCING LOSS OF TOOLS, EQUIPMENT AND MATERIALS

Have you tallied up the loss of tools and equipment in your organization lately? Naturally, it's an unpleasant subject, but let's face facts. Why not ask the accounting department to get the figures together for you? Deduct normal wear and tear, and see how much unnecessary cost you have experienced through various losses during the last year. Determine the circumstances under which each loss has occurred.

Where possible, determine the location of the theft or loss. Classify the losses accordingly; e.g., jobsite, warehouse, office, etc., and "miscellaneous undetermined." A little detective work will probably uncover some ways to reduce losses, and may lead to a suspect within the organization who will bear observation. Some of the largest thefts have often involved old and trusted employees. In one case, a long-time employee became seriously ill and his wife, in desperation, called the contractor and asked him if he would buy some of the employee's tools so that she could raise some money. The contractor went to the employee's home and found that almost every one of the tools, and a garage full of materials, were stolen over a period of many years from the contractor. The wife, evidently, was unaware of this.

Determine which thefts have occurred during off-hours. Since thefts often include materials along with tools and equipment, these should be included in the study. Was a night watchman employed? Full or part time? The cost of a night watchman is quite often much less than you would expect. One instance, for example, was a project which was under way on a busy street. The land had lain vacant prior to construction activities and a fruit and vegetable vendor had set up his little operation, selling his wares from his truck. The project manager offered to permit his activity to continue during construction with the proviso that he move into a trailer and serve as night watchman. The vendor gladly accepted.

On another project, near a university, a student was given a free place to stay in a trailer in return for night watchman services. If the student wished to be away from his duties, he arranged to get another student to take over for him. There are still a few students who are in school to gain an education (I know this is a controversial statement!), and such students enjoy the peace and quite of the construction site at night. You may even get two students who would alternate during all off-hours.

Always leave plenty of lights on at night. Lights on in the office and a radio going give the impression that a watchman is on duty. An automobile or pick-up truck left in front of the office helps convey this impression. If no watchman is on duty, an occasional visit to the site by various construction personnel helps.

Then, of course, there are the city police, sheriff, highway patrolmen, and other law enforcement agencies who can be alerted to keep an eye on the project. Some contractors engage the services of private protective patrols, but many have been found to be completely unreliable. We know of one instance where the office was entered and ransacked while the patrolman was asleep in his car out front! The burglary took place within twenty feet of the patrolman. In another instance, I came to the office to find the front door unlocked. When I confronted the patrol service, the patrolman said, "Oh, yes, I noticed the door was unlocked. I intended to call you, but never got around to it." Can you top that one?

Frequently, one of the workmen who lives in the vicinity of the project can be persuaded to make visits to the project. The visits should be made at various times without establishing a pattern.

Above all, do not wait for the big loss to occur before setting up a program of protection. Adequate protection of the premises should be your first act when moving on site.

Guard dogs are frequently used to advantage. It is best to have two dogs. In one instance, the contractor's office and yard was enclosed with chain-link fence and guarded by a dog. One thief distracted the dog while another climbed the fence on the opposite side of the yard, entered the building, and made off with considerable equipment. And, what was even worse, they ransacked the offices, scattered the files around, and really sabotaged the place. It was suspected that some disgruntled former employees were to blame, but no one was apprehended.

The construction site is a natural attraction to children. Much damage can be done to materials and equipment unless they are properly protected. For your own legal protection, always post the site conspicuously, and avoid unnecessary hazards to children.

REDUCING THE COST OF TRUCKING EQUIPMENT

Since the purchase and operating costs of over-the-road equipment constitute a substantial part of the contractor's overhead, let us consider ways to reduce these costs. Costs *can* be reduced and efficiency improved at the same time.

Obviously, the first step is to obtain the lowest purchase price on the equipment. This can be done by consolidating the buying of as many of the cars and trucks as feasible with one supplier and at one time. Try to arrange purchases at a time which suits the dealer. Order equipment well in advance. Some contractors find it advantageous to order well ahead of delivery and get new models as early in the year as possible, and then trade all equipment back at the same time each year, when it is either one or two years old. Much depends upon the extent of use and the necessity for trading off. Have a conference with the dealer and let *him* advise the time of year and the conditions which will enable you to purchase at the lowest cost. By purchasing as many vehicles as possible of the same make, simple repairs can be more easily made by your own people, and at regular scheduled maintenance periods. Weaknesses in a particular make can be anticipated and corrected before breakdown, as previously recommended.

One cost factor well worth keeping in mind is the expiration of annual licenses. It will reduce costs if equipment is traded off just prior to expiration of licenses so as to avoid buying new licenses. If equipment is traded annually, try to avoid buying licenses for two overlapping years. This is a bigger factor in some states than in others, but it should be a point of consideration.

Now we come to operating costs. Driver training is one of the biggest opportunities to reduce operating costs. Almost anyone who can walk can get a state driver's license, so that doesn't mean anything. You will have to do your own testing. A four-page standardized test, "Traffic and Driving Knowledge," is shown in Figure 9-4. Professor Neyhart, who prepared the test, is a consultant for the American Automobile Association. The administration of this test to all personnel driving company equipment will reveal the capability of drivers. Those who do not satisfactorily pass the test may then be given special driver training, or reassigned to jobs requiring little or no driving.

The test previously described is a written test. We also want to give a driving test under typical road conditions. For this purpose, see "Road Test in Traffic," (Figure 9-5). This is a two-page rating sheet for appraising the capabilities of truck drivers.

Both of these tests are available from the Dartnell Corporation (4660 Ravenswood Ave., Chicago, Ill. 60640).

The same factors which apply to the operation of other equipment apply to motor vehicle operation, such as discussed under the subject "People Create Their Own Hazards" in Chapter 10, and elsewhere in this chapter.

HOW TO GET MORE FROM YOUR TOOLS AND EQUIPMENT

Form No. 16

Standardized Test

TRAFFIC AND DRIVING KNOWLEDGE

For Drivers of Motor Trucks

Fill in these blanks:

Name_____

Street and Number_____

City_____ State_____

Age_____
Check your State law as to discrimination because of age.

Number of years of driving experience_____ Years

Possible Score—57   Letter Grade_____

Number Wrong_____

Final Score_____
(Number wrong subtracted from possible score)

Instructions to the driver: Select the one answer that you think best completes each of the following statements and questions and place a check-mark (√) before it. Skip all problems which are not clear to you on first thought and come back to them after you have finished the others. Check-mark only *one* answer in each problem. Study the following sample before you start the test:

Sample:

1. The distributor sends—

    a ( ) Oil to all the bearings
    b ( ) Gasoline to all the cylinders
    c (√) An electrical charge to each spark plug
    d ( ) Water around through the cooling system

Notice that the best answer in the sample is c, so c has been checked in the parentheses. That is the way you are to answer these questions.

DO NOT TURN THIS PAGE UNTIL TOLD TO DO SO

Prepared by Amos E. Neyhart and Helen L. Neyhart. Professor Amos E. Neyhart is Administrative Head, Institute of Public Safety, The Pennsylvania State College, State College, Pennsylvania, and Road Training Consultant, American Automobile Association, Washington, D. C.

FIGURE 9-4

Standardized Test—Traffic and Driving Knowledge

Problem
No.

1. To save gas a driver should—
    a ( ) Keep at a moderate, even speed
    b ( ) Run at high speed whenever it is safe
    c ( ) Stop the engine at all red lights
    d ( ) Coast down all hills

2. The generator—
    a ( ) Puts a hot spark at spark plugs when needed
    b ( ) Restores electrical energy in the storage battery
    c ( ) Turns the engine over to get it started
    d ( ) Generates heat for the defroster

3. You are driving in high gear and want to stop, which is the last thing you should do in stopping?—
    a ( ) Take your foot off the gas pedal
    b ( ) Step on the brake
    c ( ) Step on the clutch
    d ( ) Speed up the engine

4. Upper headlight beams should be used—
    a ( ) For city driving
    b ( ) For country driving
    c ( ) In fog
    d ( ) While passing other motor vehicles

5. The truck should be brought to a complete stop from over 30 m.p.h., under ordinary driving conditions, by stepping on the—
    a ( ) Brake pedal, then the clutch just before coming to a complete stop
    b ( ) Clutch pedal, then the brake
    c ( ) Clutch and brake pedals together
    d ( ) Brake pedal *after* putting gearshift in neutral

6. If you are involved in an accident, the *first* thing to do is to—
    a ( ) Do what you can for anyone who is hurt
    b ( ) Get the names and addresses of everyone involved, also license numbers of any cars
    c ( ) Get signed statements of witnesses
    d ( ) Call the police

7. If the radiator has frozen, it is best to—
    a ( ) Drive rapidly so the heat of the engine will thaw it out
    b ( ) Drive very slowly to the nearest place where hot water may be obtained
    c ( ) Blanket the engine and allow motor to run at idling speed until ice is thawed
    d ( ) Race engine with vehicle standing till ice is thawed

8. Which is the correct thing to do when you hear the siren of an emergency vehicle approaching, such as ambulance or fire engine?—
    a ( ) Mind your own business and continue ahead as if nothing different is happening
    b ( ) Pull over to the right curb and stop
    c ( ) Drive at extreme right to give the emergency vehicle plenty of room to pass you
    d ( ) Race after it to be of assistance if possible

9. A series of jerks when the truck is starting means—
    a ( ) Too much pressure on the gas pedal
    b ( ) Not enough pressure on the gas pedal
    c ( ) Failure to coordinate gripping pressure of the clutch with the right amount of gas pedal pressure
    d ( ) Too rich a gas mixture

Problem
No.

10. Which do you consider to be the best way to decrease accidents among commercial drivers?—
    a ( ) Posting names of drivers having accidents on company bulletin board
    b ( ) Holding weekly or monthly safety "pep" meetings
    c ( ) Giving recognition for best driving records
    d ( ) "Lecturing" drivers after each accident

11. The ammeter shows the—
    a ( ) Water level in the battery
    b ( ) Temperature of the water in the radiator
    c ( ) Rate at which the battery is being charged or discharged
    d ( ) Rate the vehicle is traveling

12. Because of natural forces pulling on the truck when you take a curve, it is best to—
    a ( ) Slow down while in the curve
    b ( ) Slow down before reaching the curve and feed gas a bit while in it
    c ( ) Apply the brakes in the curve
    d ( ) Cut a straight path across the curve

13. If the light turns amber *as you are entering* an intersection, you should—
    a ( ) Pass on through cautiously
    b ( ) Stop immediately, no matter where you are
    c ( ) Stop and back up past the cross-walk
    d ( ) Go ahead and pay no attention to the change in lights

14. At which place *may* you pass another vehicle going in the same direction?—
    a ( ) At an intersection, if there is no other traffic
    b ( ) Near a hilltop, if there is a three lane road
    c ( ) On a railroad crossing, if no train is coming
    d ( ) Between intersections, if there is no traffic rule forbidding it

15. When going down a steep hill, the good driver will—
    a ( ) Shift to lower gear after starting down grade
    b ( ) Shift to lower gear before starting down grade
    c ( ) Turn off ignition to save gas
    d ( ) Put the gears in neutral and coast

16. An intoxicating drink most seriously affects a driver's—
    a ( ) Judgment
    b ( ) Strength of grip
    c ( ) Sight
    d ( ) Reaction time

17. If you are driving into an intersection where there is no traffic control of any kind, who should be given the right-of-way?—
    a ( ) The driver on your left
    b ( ) Pedestrians
    c ( ) No one should be favored
    d ( ) You should, if you blow your horn

18. Which is the most important in becoming a *safer* driver?—
    a ( ) Having your eyes tested and wearing glasses if necessary
    b ( ) Memorizing all the rules in the vehicle code
    c ( ) Always considering the other fellow
    d ( ) Studying all the accidents that you read or hear about

19. The stopping distance of a vehicle is ordinarily affected the most by—
    a ( ) Condition of brakes
    b ( ) Tread on tires
    c ( ) Roadway surface
    d ( ) Weather conditions

FIGURE 9-4 (cont.)

HOW TO GET MORE FROM YOUR TOOLS AND EQUIPMENT 139

Problem No.

20. From the standpoint of avoiding accidents, which is the most important to keep in perfect condition at all times?—
   a ( ) Brakes
   b ( ) Rear view mirror
   c ( ) Windshield wiper
   d ( ) Tires

21. A stop sign means to come to a full stop—
   a ( ) When traffic is crossing in front of you
   b ( ) When you can't see around corners
   c ( ) During the daytime hours
   d ( ) At all times

22. The chief reason for maintaining the correct air pressure in all tires at all times is to—
   a ( ) Make the vehicle steer more easily
   b ( ) Make smoother riding
   c ( ) Keep truck from swaying when brakes are applied
   d ( ) Conserve tires by preventing tread-wear, rim bruises and carcass breakdown

23. If the driver ahead of you extends his arm straight out, you can be fairly sure he is going to—
   a ( ) Do something different
   b ( ) Turn left at the next intersection
   c ( ) Knock ashes off his cigar or cigarette
   d ( ) Pull off to the side of the road or stop

24. Which system in a truck requires more lubricant than the others?—
   a ( ) Ignition
   b ( ) Braking
   c ( ) Steering
   d ( ) Exhaust

25. Which of these should you think about to keep your attention on driving?—
   a ( ) The accidents you know about
   b ( ) The I. C. C. regulations
   c ( ) The lives and property you are responsible for
   d ( ) The company's reputation

26. What should be the good driver's opinion of a stop sign?—
   a ( ) It need not be obeyed if there is no danger
   b ( ) It is chiefly for beginners and drivers not familiar with the road
   c ( ) It is a warning to go in a lower gear
   d ( ) It means *stop* at all times

27. When starting an empty truck on packed snow or ice, it is best to use—
   a ( ) Second gear, and let the clutch pedal up slowly
   b ( ) High gear, and let the motor idle while releasing the clutch quickly
   c ( ) Low gear, and let the clutch out quickly
   d ( ) Low gear, and a somewhat higher engine speed than for ordinary circumstances

28. Traffic experts agree that—
   a ( ) Over 75% of motor vehicle accidents are caused by drivers' mistakes
   b ( ) Less than 25% are caused by drivers' mistakes
   c ( ) Half are caused by drivers and half by car defects
   d ( ) Causes of accidents are evenly divided between driver, car and road

Problem No.

29. The greatest danger from carbon monoxide is in—
   a ( ) Traveling too closely behind the vehicle ahead
   b ( ) Running the engine in a closed building
   c ( ) Failing to have fresh air in the cab at all times while driving
   d ( ) Running the engine when the truck is standing still

30. What should you do when stopping on an upgrade waiting for the red light to change to green?—
   a ( ) Race the motor to be sure it won't stall and use hand brake
   b ( ) Put gears in neutral and use the foot brake
   c ( ) Slip the clutch to keep from rolling back
   d ( ) Put gears in neutral and use the hand brake

31. If you find yourself driving on soft, muddy ground, the best thing to do is—
   a ( ) Stop and try to start out in low gear
   b ( ) Slow down
   c ( ) Keep going, change to a lower gear, if necessary
   d ( ) Stop, put in reverse and back out

32. The best way to save brakes on a downgrade is to—
   a ( ) Set hand brake about half way
   b ( ) Shift to a lower gear before starting downgrade
   c ( ) Keep brakes on lightly
   d ( ) Use brakes only for real emergencies

33. When may you drive on the left side of a two-lane road?—
   a ( ) While rounding a sharp curve to the left
   b ( ) When the lights of a car close behind annoy you
   c ( ) While passing the vehicle ahead on the straightaway
   d ( ) When a trolley has stopped to let off and take on passengers

34. Which one do you consider most important in a *uniform* traffic code?—
   a ( ) Shape and position of traffic signs
   b ( ) Hand and mechanical signals
   c ( ) Lighting regulations (as, no headlights when fog lights are used)
   d ( ) Parking regulations

35. Only in emergencies does the good driver use his—
   a ( ) Horn
   b ( ) Brakes
   c ( ) Low beam lights
   d ( ) Rear view mirror

36. Brakes should always be tested by driver—
   a ( ) After a long trip
   b ( ) Before starting daily operation
   c ( ) Before putting the vehicle away at night
   d ( ) Each time the tires are inflated

37. The highest number of traffic deaths is among—
   a ( ) Drivers
   b ( ) Passengers
   c ( ) Pedestrians
   d ( ) Bicyclists

38. If you have a breakdown in the daytime on the highway and are alone, which would you do?—
   a ( ) Walk to the nearest telephone
   b ( ) Hitch-hike to nearest telephone
   c ( ) Ask some passer-by to stay with vehicle while you go to call
   d ( ) Ask a passer-by to make the call for you

FIGURE 9-4 (cont.)

Problem No.

39. A driver can best avoid trouble by—
   a ( ) Driving slower than other traffic
   b ( ) Keeping a foot lightly on the brake
   c ( ) Obeying all the signs and signals
   d ( ) Recognizing trouble in the making—defensive driving

40. An extra rear-view mirror on the right side of a vehicle will help compensate or make up for—
   a ( ) Poor hearing
   b ( ) Poor eyesight
   c ( ) Tiredness and inattention
   d ( ) Slow reaction time

41. Which of these is the most important for all weather safe driving?—
   a ( ) Defroster
   b ( ) Heater
   c ( ) Windshield wiper
   d ( ) Flashlight

42. While learning to drive, which method would help a driver progress faster?—
   a ( ) Experimenting with the truck by himself
   b ( ) Practicing under a trained instructor
   c ( ) Reading it all from a book
   d ( ) Asking all his friends how they do the things he is learning

43. In most situations, who should be favored if there is no traffic light or officer to direct traffic?—
   a ( ) Pedestrians at all times
   b ( ) Pedestrians, only if they are already in the crosswalk
   c ( ) Motorists at all times, because they cannot stop quickly
   d ( ) Motorists, if they give a warning horn blow

44. Which drivers, considering driving population, are the most unfit?—
   a ( ) The partially deaf
   b ( ) The totally deaf
   c ( ) The color blind
   d ( ) Those with only one arm

45. When you meet glaring lights at night what should you *not* do?—
   a ( ) Keep your eyes on the oncoming lights
   b ( ) Look at the right side of the road
   c ( ) Look straight ahead
   d ( ) Look out the corner of your eyes

46. After driving through fairly deep water what part of a truck is most likely to fail to work immediately?—
   a ( ) Steering mechanism
   b ( ) Braking system
   c ( ) Lighting system
   d ( ) Engine

47. Accident repeaters are generally drivers who—
   a ( ) Have poor vision (eyes)
   b ( ) Don't know traffic rules
   c ( ) Have slower than average reaction time
   d ( ) Have never developed real driving skill

48. Which of the following do you consider would bring the greatest reduction in accidents?—
   a ( ) Construction of "fool-proof" highways
   b ( ) Mechanical perfection of the vehicle
   c ( ) Installation of needed signals, signs and markings
   d ( ) Training of all beginning drivers by trained teachers

Published by The Dartnell Corporation, Chicago, Ill. 60640
Printed in U.S.A.

Problem No.

49. Under what conditions should you resort to "ticket fixing"?—
   a ( ) If you did not know you had broken a rule and yet had gotten a ticket
   b ( ) If a good friend of yours was in a position to "fix" your ticket without anyone knowing it
   c ( ) If you would lose your job if the boss found out you had received a ticket
   d ( ) Under no conditions

50. Which group do you think have the greatest percentage of accidents for the number of miles they drive?—
   a ( ) Truck drivers
   b ( ) Bus drivers
   c ( ) Passenger car drivers
   d ( ) No one group

51. When parking a truck on an upgrade, the best practice is to—
   a ( ) Get close to the curb and turn the front wheels away from curb
   b ( ) Turn the front wheels toward the curb
   c ( ) Put the gearshift in neutral
   d ( ) Put gearshift in high

52. Which neglected part do you consider to be the most dangerous?—
   a ( ) One headlight out
   b ( ) Missing engine
   c ( ) Cracked windshield
   d ( ) Horn that doesn't work

53. What is the best way you can show the driver behind you that you are going to turn at an intersection?—
   a ( ) Straddle lanes so no one can pass you
   b ( ) Blink your stop lights
   c ( ) Slow down far ahead of intersection
   d ( ) Place your truck in the correct lane for the turn you are going to make

54. The oil gauge on the dashboard tells—
   a ( ) If the oil filter is working properly
   b ( ) The amount of oil reserve
   c ( ) The amount of oil being used
   d ( ) The pressure at which oil is being pumped

55. What is the chief purpose of a state law requiring a driver to show that he can handle a truck before receiving a license to drive?—
   a ( ) To provide a reference list of drivers
   b ( ) To produce highway funds
   c ( ) To be able to track down criminals
   d ( ) To determine the fitness of people to drive

56. A good way to start most trucks in cold weather is to—
   a ( ) Step on the starter before turning on ignition
   b ( ) Put gearshift in neutral but don't depress clutch
   c ( ) Depress the clutch with gearshift in neutral
   d ( ) Release the hand brake before turning on ignition

57. The chief cause of skidding is—
   a ( ) Snow and ice on the road
   b ( ) Brakes improperly adjusted
   c ( ) Tires too smooth
   d ( ) Driving too fast for conditions

FIGURE 9-4 (cont.)

Form No. 18

Check List and Score Sheet for

ROAD TEST IN TRAFFIC

for

Testing, Selecting, Rating and Training Truck Drivers

By Amos E. Neyhart

The material used herein is based on copyrighted material of the Institute of Public Safety and American Automobile Association and reproduced with the permission of the association and the author, Amos E. Neyhart, Administrative Head, Institute of Public Safety, The Pennsylvania State College and Consultant on Road Training, American Automobile Association.

Name of Driver _____ Final Score (Sum of Parts I and II) _____

Street and Number _____ Final Letter Grade _____

City and State _____ Date _____

PART I — SPECIFIC

ITEMS	DEDUCT	CHECK (√) ITEMS MISSED BY DRIVER	DEDUCTIONS

I. CHECKING THE DRIVER

A. Fails to enter vehicle from curb side—when practical — 2
B. Fails to check doors to see if closed properly — 2
C. Fails to adjust windows for ventilation — 2
D. Fails to adjust rear-view mirrors — 3
E. Fails to adjust seat properly — 1
F. Fails to assume erect and alert driving position — 1

II. STARTING ENGINE

A. Fails to depress clutch pedal — 1
B. Does not check gearshift lever for neutral position — 2
C. Fails to turn on ignition switch before pressing starter button — 1
D. Does not release starter button as soon as engine starts to operate on its own power — 2
E. Spends too much time trying to get engine to run, fails to use choke properly — 1
F. Does not allow engine to warm up — 5
G. Races engine during warm-up period — 5
H. Fails to check air pressure — 5

III. STARTING THE VEHICLE IN LOW

A. Fails to check traffic conditions — 5
B. Selects wrong gear (does not start in low) — 3
C. Does not release hand brake — 1
D. Rolls back when on a grade — 5
E. Races the engine — 5
F. Stalls the engine — 5

IV. BACKING

A. Fails to stop in correct position to back — 5
B. Fails to go to rear of vehicle before backing — 5
C. Fails to use both mirrors when backing — 5
D. Fails to keep guide in sight — 3
E. Backs jerkily — 2
F. Oversteers and zigzags when backing — 2

V. CLUTCHING, SHIFTING GEARS

A. Rides the clutch — 3
B. Fails to keep eyes on the road during shifting maneuver — 3
C. Stays in low gear(s) too long — 3
D. Fails to attain proper speed when shifting to higher gears — 3
E. Stays in high gear(s) too long — 3
F. Stalls the engine — 5
G. Fails to "double clutch" and clashes gears (any other clashing of gears) — 5
H. Slips clutch to hold vehicle from rolling back while waiting at traffic signal — 3
I. Keeps clutch pedal depressed while waiting at traffic signal — 1
J. Selects wrong gear—upgrade, downgrade or on level — 3
K. Coasts down grades, up to stop signs and traffic lights — 3

VI. STEERING

A. Places hands in unstable position on wheel — 2
B. Steers abruptly, not smoothly — 5
C. Rests arm on window — 2
D. Uses one hand occasionally — 1
E. Turns steering wheel while vehicle is at rest — 2

VII. RAILROAD CROSSING

A. Fails to look in all directions — 5
B. Fails to come to full stop when necessary — 5
C. Fails to stop at a safe place, if necessary — 5
D. After stopping fails to shift to lower gear and remain in that gear until clear of tracks — 5
E. Fails to drive in correct position when crossing tracks — 5

VIII. SPEED CONTROL (Exclusive of Turns)

A. Too fast for conditions — 5
B. In excess of marked speed limits — 4
C. Too slow for conditions — 2
D. Brakes on curves — 5

FIGURE 9-5

Check List and Score Sheet for Road Test in Traffic

(Dartnell)

142 HOW TO GET MORE FROM YOUR TOOLS AND EQUIPMENT

ITEMS	DEDUCT	CHECK (√) ITEMS MISSED BY DRIVER	DEDUCTIONS
IX. STOPPING			
A. Before necessary (especially at signals and signs)	1	☐	
B. Not soon enough (over-running crosswalk or avoidance zone line)	2	☐	
C. Not at a safe place (too close to other vehicles, etc.)	5	☐	
X. STOP STREETS			
A. Fails to come to full stop	5	☐	
B. Fails to stop in a position to see roadway to the right and left (second stop if necessary)	5	☐	
C. Hesitates too long for conditions	3	☐	
XI. UNCONTROLLED INTERSECTIONS OR THROUGH STREETS			
A. Fails to slow down (to stop if necessary)	3	☐	
B. Fails to look in all directions	5	☐	
C. Fails to shift to lower gears when necessary	3	☐	
D. Fails to respond to hazardous traffic conditions in the making	5	☐	
XII. SIGNALING FAILURES			
A. Leaving curb—fails to signal	2	☐	
B. Leaving curb—fails to look back	2	☐	
C. Turning—fails to use turn signals	2	☐	
D. Leaves turn signal on, after turning	2	☐	
E. Does not use turn signals moving from lane to lane	2	☐	
F. Uses horn improperly or fails to use horn	2	☐	
G. Fails to observe courtesy of signaling—hand signals when possible	5	☐	
XIII. SIGNAL VIOLATIONS			
A. Traffic signal (through on amber)	3	☐	
B. Traffic signal (through on red)	5	☐	
C. Traffic officer	5	☐	
XIV. PASSING OTHER VEHICLES GOING IN SAME DIRECTION			
A. Fails to make sure road ahead and behind is clear	5	☐	
B. Misjudges speed of oncoming traffic	5	☐	
C. Passes on curve	3	☐	
D. Passes at intersection	3	☐	
E. Passes at crest of hill	5	☐	
F. Cuts back into line too soon after passing	3	☐	
G. Passes by weaving through traffic	5	☐	
H. Starts passing when approaching obstructions in center of street	5	☐	
I. Pulls into center traffic lane when approaching center of street obstructions such as pedestrian islands	5	☐	
J. Passes so as to block vehicles of right from steering around parked or slow moving vehicle	5	☐	
K. Fails to observe indications that parked vehicle may start from curb	3	☐	

Published by The Dartnell Corporation, Chicago, Ill. 60640   Printed in U.S.A.

ITEMS	DEDUCT	CHECK (√) ITEMS MISSED BY DRIVER	DEDUCTIONS
XV. POSITION OF VEHICLE ON ROADWAY			
TRAFFIC LANES (Exclusive of Turns—Marked or Unmarked)			
A. Fails to drive in proper lane	5	☐	
B. Straddles traffic lanes (marked or unmarked)	5	☐	
C. Straddles at signal or sign when stopping	5	☐	
D. Follows too close to other vehicles	5	☐	
E. Drives too close to other vehicles, moving objects, etc.	3	☐	
TURNING (Right)			
A. Approaches from improper lane	2	☐	
B. At improper speed (too fast or too slow)	2	☐	
C. In improper lane during turn	3	☐	
D. Into improper lane after turn	3	☐	
E. Strikes curb	3	☐	
F. Makes turn unnecessarily wide	1	☐	
G. Shies away, then turns right	2	☐	
H. Shifts gears while turning	2	☐	
TURNING (Left)			
A. Approaches from improper lane	3	☐	
B. At improper speed (too fast or too slow)	2	☐	
C. In improper lane during turn	3	☐	
D. Into improper lane after turn	3	☐	
E. Cuts corner too short	1	☐	
F. Cuts corner too wide	2	☐	
G. Shies away, then turns left	2	☐	
H. Shifts gears while turning	2	☐	
XVI. SMOOTHNESS OF OPERATION			
A. Rough starts—By Jerk Recorder or Tumbling Cylinders Tally	(Deduct one point each time cylinder is tipped)		
B. Rough stops—By Jerk Recorder or Tumbling Cylinders Tally	(Maximum total deduction—10 points)		
C. Uses clutch roughly	5	☐	
D. Uses brakes roughly or unevenly	5	☐	
E. Fails to hold accelerator steady	5	☐	

PART II—GENERAL

ITEMS	0 Not at all	5 Occasionally	10 Part of time	15 Often	20 Over entire route	TOTAL DEDUCTIONS
I. Inattentive (day dreams, etc.)	•	•	•	•	•	
II. Nervous and Hesitant	0 Not at all	5 Occasionally	10 Part of time	10 Often	15 All the time	
III. Overconfident	0 Not at all	•	5 Part of time	•	10 Cocky	
IV. Fails to USE rear-view mirrors	0 Not at all	•	2.5 Part of time	•	5 Over entire route	
V. Fails to anticipate or respond to hazardous traffic conditions in the making (including pedestrians)	0 Not at all	•	5 Part of time	•	10 All the time	

PART I—TOTAL SCORE _____

PART II—TOTAL SCORE _____

GRAND TOTAL SCORE _____

Checker's Signature _____

FIGURE 9-5 (cont.)

The next step in lowering costs of over-the-road equipment is the establishment of good preventive maintenance procedures as described at the beginning of this chapter. If the foregoing steps are accomplished effectively, there will be a noticeable decrease in the cost of your vehicular equipment.

OPERATOR AND CRAFT SCHOOLS

In July, 1969, Western School of Heavy Equipment Operation sold its assets and organization to the Associated General Contractors of America, Idaho Branch, Inc., and the International Union of Operating Engineers, Local 370. The operation of the school has continued under the jointly administered training trust. A Five Craft agreement has been negotiated, including operating engineers, carpenters, teamsters, laborers, and cement masons. This unique institution, located in Weiser, Idaho, combines the skill and knowledge of management and labor to provide a permanent facility and program to train men for the construction industry.

The school was founded as a cooperative venture by management and labor to help alleviate the growing shortage of trained people needed by the industry. It is designed to upgrade or improve skills as technology advances, and to provide a training facility where individuals without prior skills or training can be given a chance to enter the industry. The school is G.I. approved.

For further information, contact AGC-Operating Engineers Training School, P.O. Box 510, Weiser, Idaho 83672.

Another school of heavy equipment operation is the Universal Heavy Construction Schools, 1901 N.W. 7th St., Miami, Florida 33125. The training consists of a sixty-lesson preparatory program which is completed at home by the student. This is followed by five weeks of training at the school's resident training facility in Homestead, Florida. Here the student is given both classroom work and actual operation of numerous types of heavy construction equipment. An employment guidance and advisory service is provided for the graduates of the school.

No doubt, training at one of these operator schools equips the student with considerable valuable knowledge of construction equipment; however, most of the graduates have little or no previous construction experience. Contractors must realize that these operators have plenty to learn about actual field operations. Much depends upon the individual's capability to absorb the brief training and translate it into practical terms. Although additional training may be required by the contractor's own personnel, the background provided by an operator school will greatly facilitate the operator's grasp of your own work. The training is G.I. approved.

USE OF HELICOPTERS IN CONSTRUCTION

Although contractors are familiar with the application of helicopters for lifting and placing loads in construction work, many additional jobs can be done faster and easier by helicopter. It is often advantageous to move and place bulky mechanical equipment in place in the basement or first floor before the building becomes closed in with the structure. Consider using helicopter pick-up and delivery all in one swift operation. Also try to consolidate all the possible jobs to be done at the same time (such as relocation of materials to a more advantageous position in or near the point of use).

Even partitions and furnishings can often be hoisted into windows to save handling time. Sometimes bulky pieces can be moved in through windows, saving disassembly and re-assembly costs. Large plate glass has been lifted to the proper floor for installation, eliminating the necessity of erecting a gin pole, or moving in a crane. Frequently helicopters prove to be advantageous in congested areas with insufficient space for cranes or other hoisting equipment to operate.

One contractor purchased a helicopter and has contracted outside work for it. He claims it has been one of the most profitable pieces of equipment he owns.

APPLICATION OF THE LASER BEAM TO CONTRACTING

The laser beam is finding new uses in building and construction work every day. Several models are in use of almost every type of job, and the applications seem to be limitless. The instrument produces a pencil-thin, harmless (low-intensity) beam of red laser light that can be projected up, down, or horizontally from the instrument.

Contractors are using the laser instrument to erect scaffolding, plumb chimneys, plumb columns and forms, erect suspended ceilings, and for just about any alignment requirement. Laser transits now replace the conventional transit for numerous applications.

The big advantage, as I see it, is that the instrument can be used by one man alone, whereas the transit requires two men. Then, too, it is much faster in many applications.

Additional information can be obtained from Spectra-Physics, 1250 West Middlefield Road, Mountain View, Calif. 94040.

THE CLAW

The "claw" is a ripping and cutting attachment which replaces the bucket on a backhoe. Uses for this unusual device are numerous. It can be used to cut off roots or limbs, break out curbing, or load trees and other debris. It can be used to cut a utility trench. It has a scissor-like claw which has unique capability. This comparatively new type of equipment may ultimately find many uses in construction operations. More information on this can be obtained from Construction Technology, Inc., P.O. Box 218, Arlington, Tex. 76010.

UNDERGROUND PIERCING EQUIPMENT

Placing pipe or utility lines under driveways, sidewalks, through embankments or other inaccessible places can be done without excavating by the use of drilling or piercing equipment. Holes a few inches or several feet in diameter can be bored up to 200 feet or more.

HOW TO BREAK UP BOULDERS AND CONCRETE

The traditional method of breaking up boulders and large masses of concrete has been by demolition, but there are situations where this method is objectionable, if not prohibitive. In the first place, demolition must be handled by professionals and specialists. It is suicide for the novice. There are methods, however, which can be effectively utilized by the layman. One of these is the splitter. The work is done by drilling a hole as is done for blasting or demolition. The splitter has a wedge-shaped probe which is inserted into this hole. Then the split wedges are moved under hydraulic or other power so that the split probe expands. This exerts great internal pressure and the rock mass fractures. Pressures of 800,000 pounds or more can be introduced within the boulder, enough to split it.

Splitters are made by Emaco, Inc., 111 Van Riper Ave., East Paterson, N.J. 07407; and by Kelly Industrial Company, Beresford, S.D. 57004.

IMPROVING YOUR WOODWORKING OPERATIONS

Since wood is one of our basic construction materials, consideration should be given to the efficient cutting and working of lumber. The most frequent woodworking operation, obviously, is cutting materials to length. The cross-cut operations include straight cutoff, miter, compound miter, bevel cutoff, and other variations or modifications.

The radial saw is generally used for these operations, and there are several important points which I would like to bring out:

1. On most of the radial saw installations I have seen on construction projects, the saw was producing a ragged cut due to misalignment of the saw blade with the line of travel. In other words, the saw blade was not traveling in a plane *parallel to the travel of the carriage.* This is an important point, because it is the biggest source of trouble with radial saw operations. Many saw operators are unaware of the problem and do not know how to properly make this adjustment. Every radial saw should be checked and adjusted if necessary at least weekly during use. This is a simple procedure and is checked as follows: Place a block of wood two to three inches thick firmly on the table of the saw up against the guide strip and just to the left of the saw blade. Now pull the carriage out so that the saw blade is just forward of the edge of the block. Place a pencil on the block so that it just touches the saw blade and hold it firmly while you pull the saw forward to a point at the rear of the blade. Rotate the blade around so that the pencil point is at the same location on the blade as it was at the starting position. If the saw blade is parallel to the line of travel, the pencil will make contact both in the front and rear positions of the carriage; otherwise, it must be adjusted. Follow the manufacturer's instructions for realignment.

Although methods of checking are given in the operating instructions of most radial saws, the method given is not always the best. I have aligned many radial saws by this method.

Failure to maintain this important adjustment will result in rough cuts, overworking the motor, short saw blade life, inaccuracies, and other complications. This adjustment is especially important when using hollow ground blades.

2. Set up the radial saw with roller tables on both sides of the saw table, rigidly connected, and in the same horizontal plane as the table of the saw. Each time the saw is relocated, all adjustments should be checked and reset if necessary. Be sure the operators know the adjustments, and place one person in charge of maintenance. In setting up the radial saw, the arm and table should slope slightly lower at the back side to prevent the carriage from creeping forward while the saw is running. Some machines have automatic returns, but even then it is best to provide a slight slope about one-eighth inch lower at the back.

3. On most jobs, radial saws are allowed to accumulate sawdust within the arm and pack dust and dirt on the rollers. This not only makes the machine difficult to operate, but it also causes a very high rate of wear on parts which are expensive to replace. Excessive wear on these vital parts causes a loss of accuracy and no adjustment is possible. *The inside of the arm and the rollers should be thoroughly cleaned daily.* This requires only about five minutes' time and it is repaid several times each day in less operating effort.

4. Be sure the power supply is adequate. Don't make the common mistake of using long extension cords of too-light gauge. The heavier the wire, the less the voltage drop. If in doubt, have the electrician measure the voltage at the saw *while under load.* If power supply is still low after proper wiring, call the power company and have them check the supply.

Electrical motors will overheat if operated on too-low voltage. Motor efficiency drops rapidly as voltage decreases. I have seen many saw installations where 5 HP motors were delivering no more than 3 HP due to excessive voltage drop or too low supply.

5. Set up a saw shed to protect the saw from the weather. It is a good idea to have at least one box about three feet in length to take shorts, and another for scraps. The work area should be kept clean and materials assorted into sizes, otherwise there will be unnecessary waste of materials.

6. Saws, and especially radial saws, are very dangerous in the hand of inexperienced operators. New operators should be given very careful and complete training by a capable operator. Additional information will be found in Chapter 10 pertaining to the safe operation of woodworking machines.

10

Complying with the Occupational Health and Safety Requirements

The thirty-two page federal government pamphlet in front of me is simply entitled, "An Act." But beneath this simple title lies a tremendous burden for contractors, as if the contractor didn't have enough burdens already. This particular Act, better known as the "Occupational Safety and Health Act of 1970," has created a whole new field of headaches. The purpose, to quote directly from the Act, is "to assure safe and healthful working conditions for working men and women; by authorizing enforcement of the standards developed under the Act; by assisting and encouraging the States in their efforts to assure safe and healthful working conditions; by providing for research, information, education, and training in the field of occupational safety and health; and for other purposes."

WHO IS AFFECTED BY THE ACT?

Quoting from a U.S. Department of Labor pamphlet on recordkeeping requirements, "The provisions of the law apply to every employer engaged in business affecting commerce who has employees." This makes it an entirely different ball game from the first passage of the Act which covered all those engaged in *interstate* commerce. By dropping the "interstate" definition, it now seems that every business will come under the law.

Since my first study and digest of the law, it has been revised, expanded, clarified in places, relaxed in other places, and complicated in other places. For the past several months I have been reviewing these many changes and they are still coming through.

The general industry standards, the construction standards, the compliance operations manual and various other government-issued documents, plus a complete explanation of regulations, are available from Prentice-Hall, Inc., in their Labor Relations Guide. Further information on this valuable loose-leaf service may be obtained by writing to Prentice-Hall, Inc., Englewood Cliffs, N.J. 07632. Attention: Service Sales.

The Act has given the Secretary of Labor broad and almost unlimited powers in setting *additional* safety and health standards, so there is no end in sight on the extent of the new law. The Act authorizes the Secretary of Labor to promulgate as occupational safety and health standards any existing Federal Standards or any of the national standards (such as those issued by the National Fire Protection Association). The Secretary had these powers until April 28, 1973, which gave him time enough to "promulgate" about everything under the sun into law!

In addition, the Secretary of Labor may, upon basis of information submitted by HEW, advisory committees and others, revise, modify, or revoke existing standards as well as promulgate new ones. This can be used to our advantage if contractor associations will build up the pressure on the Secretary of Labor to revise many of the regulations and revoke many more. Since many of these regulations do not actually improve safety, certainly the logic of modifying, revising or revoking them makes sense in view of the fact that they serve only to increase construction costs.

COMPLAINTS OF VIOLATIONS

Any employee or his representative may request an inspection by the Department of Labor. This means that any disgruntled employee may cause endless problems by filing a complaint. He can always find some violation. I doubt if there is a contractor in the country who is in full compliance, even after making a concerted effort to comply.

ENFORCEMENT

Safety inspectors from the Labor Department "may enter without delay, and at any reasonable times, any establishment covered by the Act to inspect the premises and all pertinent conditions, structures, machines, apparatus, devices, equipment, and materials therein, and to question privately any employer, owner, agent or employee."

When an investigation reveals a violation, the employer is issued a written citation. The citation will fix a time for abatement of the violation and a copy must be posted prominently near the place where the violation occurred. Then the Labor Department will notify the employer of the penalty, if any, which will be assessed.

Of course, there is an appeal procedure, but most likely it will be less expensive to pay the fine than fight the thing back and forth for an undetermined length of time.

PENALTIES FOR VIOLATIONS

It is expected that penalties for first offenses will not be severe; however, for repeated violations penalties can be assessed up to $10,000 *for each violation*. This law has teeth in it, Mr. Contractor! "Any employer who fails to correct a violation for which a citation has been issued within the period of time prescribed therein may be penalized up to $1,000 *each day* the violation persists." After that it gets rougher. "A willful violation by an employer which results in the death of any employee is punishable by a fine of up to $10,000 or imprisonment for up to six months."

RECORDS YOU MUST KEEP

Employers are required to maintain accurate records, and periodic reports, of illnesses, injuries and especially deaths resulting from work environment factors. Although minor injuries need not be recorded, a record must be kept if it involves "medical treatment, loss of consciousness, restriction of work or motion, or transfer to another job."

Each recordable occupational injury and occupational illness must be entered on a log of cases (OSHA Form No. 100 shown in Figure 10-1) within two working days of receiving information that a recordable case has occurred. Logs must be kept current and retained for five (5) years following the end of the calendar year to which they relate.

Logs are to be maintained for three purposes:

1. Logs for the prior five (5) year period must be available in the establishment without delay and at reasonable time for examination by representatives of the Department of Labor or the Department of Health, Education and Welfare, or States accorded jurisdiction under the Act.

2. The log will be used in preparing the annual summary of Occupational Injuries and Illnesses (OSHA Form No. 102, shown in Figure 10-2) which must be posted in every establishment.

3. Those establishments selected to participate in a statistical program will be required to prepare a report based on entries in this log.

In addition, the log will aid you in reviewing the occupational injury and illness experience of your employees.

To supplement the Log of Occupational Injuries and Illnesses (OSHA No. 100), each establishment must maintain a record of each recordable occupational injury or illness. Workmen's compensation, insurance, or other reports are acceptable as records if they contain all facts listed below or are supplemented to do so. If no suitable report is made for other purposes, this form (OSHA No. 101, shown in Figure 10-3) may be used or the necessary facts can be listed on a separate plain sheet of paper. These records must also be available in the establishment without delay and at reasonable times for examination by representatives of the Department of Labor and the Department of Health, Education and Welfare, and States accorded jurisdiction under the Act. The records must be maintained for a period of not less than five years following the end of the calendar year to which they relate. However, starting January 1, 1973, employers of seven (7) or fewer employees won't have to maintain the log of occupational injuries and illnesses (OSHA Form No. 100), supplementary records (OSHA Form

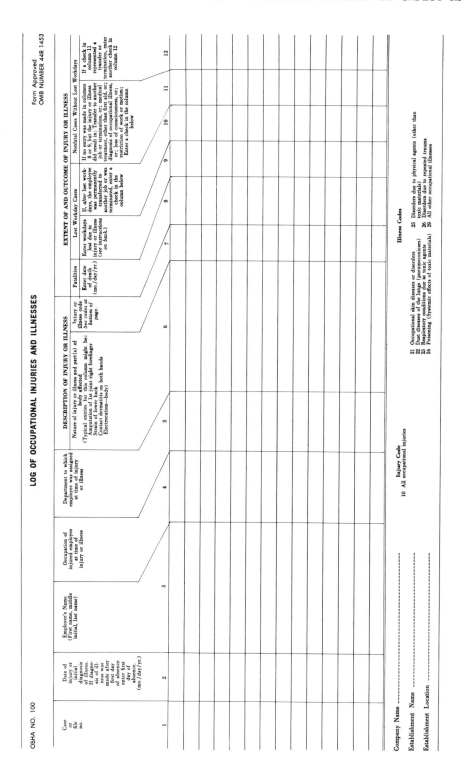

FIGURE 10-1

Log of Occupational Injuries and Illnesses

COMPLYING WITH THE OCCUPATIONAL HEALTH AND SAFETY REQUIREMENTS

OSHA No. 102

Form Approved
OMB No. 44R 1453

Summary

Occupational Injuries and Illnesses

Establishment Name and Address:

Injury and Illness Category		Fatalities	Lost Workday Cases			Nonfatal Cases Without Lost Workdays*	
			Number of Cases	Number of Cases Involving Permanent Transfer to Another Job or Termination of Employment	Number of Lost Workdays	Number of Cases	Number of Cases Involving Transfer to Another Job or Termination of Employment
Code 1	Category 2	3	4	5	6	7	8
10	Occupational Injuries						
	Occupational Illnesses						
21	Occupational Skin Diseases or Disorders						
22	Dust diseases of the lungs (pneumoconioses)						
23	Respiratory conditions due to toxic agents						
24	Poisoning (systemic effects of toxic materials)						
25	Disorders due to physical agents (other than toxic materials)						
26	Disorders due to repeated trauma						
29	All other occupational illnesses						
	Total—occupational illnesses (21-29)						
	Total—occupational injuries and illnesses						

*Nonfatal Cases Without Lost Workdays—Cases resulting in: Medical treatment beyond first aid, diagnosis of occupational illness, loss of consciousness, restriction of work or motion, or transfer to another job (without lost workdays).

FIGURE 10-2

Summary—Occupational Injuries and Illnesses

OSHA No. 101
Case or File No. _____

Form approved
OMB No. 44R 1453

Supplementary Record of Occupational Injuries and Illnesses

EMPLOYER
1. Name _____
2. Mail address _____
   (No. and street)   (City or town)   (State)
3. Location, if different from mail address _____

INJURED OR ILL EMPLOYEE
4. Name _____ Social Security No. _____
   (First name)   (Middle name)   (Last name)
5. Home address _____
   (No. and street)   (City or town)   (State)
6. Age _____   7. Sex: Male _____ Female _____ (Check one)
8. Occupation _____
   (Enter regular job title, *not* the specific activity he was performing at time of injury.)
9. Department _____
   (Enter name of department or division in which the injured person is regularly employed, even though he may have been temporarily working in another department at the time of injury.)

THE ACCIDENT OR EXPOSURE TO OCCUPATIONAL ILLNESS
10. Place of accident or exposure _____
    (No. and street)   (City or town)   (State)
    If accident or exposure occurred on employer's premises, give address of plant or establishment in which it occurred. Do not indicate department or division within the plant or establishment. If accident occurred outside employer's premises at an identifiable address, give that address. If it occurred on a public highway or at any other place which cannot be identified by number and street, please provide place references locating the place of injury as accurately as possible.
11. Was place of accident or exposure on employer's premises? _____ (Yes or No)
12. What was the employee doing when injured? _____
    (Be specific. If he was using tools or equipment or handling material, name them and tell what he was doing with them.)
13. How did the accident occur? _____
    (Describe fully the events which resulted in the injury or occupational illness. Tell what happened and how it happened. Name any objects or substances involved and tell how they were involved. Give full details on all factors which led or contributed to the accident. Use separate sheet for additional space.)

OCCUPATIONAL INJURY OR OCCUPATIONAL ILLNESS
14. Describe the injury or illness in detail and indicate the part of body affected. _____
    (e.g.: amputation of right index finger at second joint; fracture of ribs; lead poisoning; dermatitis of left hand, etc.)
15. Name the object or substance which directly injured the employee. (For example, the machine or thing he struck against or which struck him; the vapor or poison he inhaled or swallowed; the chemical or radiation which irritated his skin; or in cases of strains, hernias, etc., the thing he was lifting, pulling, etc.)
_____
16. Date of injury or initial diagnosis of occupational illness _____
    (Date)
17. Did employee die? _____ (Yes or No)

OTHER
18. Name and address of physician _____
19. If hospitalized, name and address of hospital _____

Date of report _____ Prepared by _____
Official position _____

FIGURE 10-3

Supplementary Record of Occupational Injuries and Illnesses

No. 101), or the annual summary (OSHA Form No. 103). All employers are still required to report accidents involving a fatality or the hospitalization of five or more employees within 48 hours.

Such records must contain at least the following facts:

1. *About the employer:* name, mail address, and location if different from mail address.
2. *About the injured or ill employee:* name, social security number, home address, age, sex, occupation, and department.
3. *About the accident or exposure to occupational illness:* place of accident or exposure, whether it was on employer's premises, what the employee was doing when injured, and how the accident occurred.
4. *About the occupational injury or illness:* description of the injury or illness, including part of body affected; name of the object or substance which directly injured the employee; and date of injury or diagnosis of illness.
5. *Other:* name and address of physician; if hospitalized, name and address of hospital; date of report; and the name and position of person preparing the report.

The official OSHA poster must be displayed at the workplace.

EMPLOYERS' OBLIGATION TO EMPLOYEES

Employers will be required to keep their employees informed of their protections and obligations by the posting of notices "or other appropriate means."

NEW ASSISTANCE FROM THE SMALL BUSINESS ADMINISTRATION

The Act also provides for financial assistance through the SBA for alterations to equipment, facilities, or other methods of operation to comply with the regulations. The SBA will provide assistance based upon their determination of the need in each instance.

OSHA PUBLICATIONS

Following is a list of the publications and forms available on February 18, 1972. These should be available from your nearest office of the U.S. Department of Labor, Occupational Safety and Health Administration. They are also available from the U.S. Government Printing Office, Documents Department, Washington, D.C. 20402. My recent experience, however, is that the printing office required over two months to answer my inquiry and, even then sent me only a list of prices of the publications and a form letter explaining how to order them.

1. Compliance Operations Manual, A Manual of Guidelines for Implementing the Occupational Safety and Health Act of 1970, Jan. 1972.
L 35.8:C 73   S/N 2916-0006 $2.00
2. Handy Reference Guide, Williams-Steiger Occupational Safety and Health Act of 1970.
L 35.8:W 67   S/N 2915-0001   .20

3. Inspection Survey Guide, A Handbook of Guides and References to Safety and Health Standards for Federal Contracts Programs.
L 16.3:326   S/N 2903-0111 $2.25

4. Occupational Safety and Health Act of 1970. An Act to Assure Safe and Healthful Working Conditions for Working Men and Women; by Authorizing Enforcement of the Standards Developed Under the Act; by Assisting and Encouraging the States in Their Efforts to Assure Safe and Healthful Working Conditions; by Providing for Research, Information, Education, and Training in the Field of Occupational Safety and Health; and for Other Purposes. Approved Dec. 29, 1970.
91-2:Pub.Law 596   .20

5. Occupational Safety and Health Standards; National Concensus Standards and Established Federal Standards, Federal Register, Vol. 37, No. 203, October 19, 1972.   .20

6. Safety and Health Regulations for Construction with Pertinent Amendments, Reprinted from the Federal Register.
L 35.6:C 76   S/N 2915-0009   .70

7. Recordkeeping Requirements under the OSHA of 1970.
USGPO:1971 O-429-570   n/c

Notice that part of the objective of the Act is to "encourage the states in their efforts to assure safe and healthful working conditions." This includes monetary grants covering up to 50 percent of the cost of the program. Many of the states are already working up their own sets of regulations and requirements. Several of the states already have rigid safety and health laws. Others which have had laws on the books are now preparing to enforce them. All of this adds up to the fact that everyone in the construction industry must learn and apply these safety and health regulations whether we like them or not.

It is not feasible to quote the entire law here as it would occupy the entire book, but I will digest the salient features in an effort to convey the seriousness and extensive nature of the law. You can be sure that the coverage *is* extensive and, in the opinion of most industry people, entirely too tight.

Paragraph 1926.10 of the previously cited document states that "no contractor or subcontractor contracting any part of the contract work shall require any laborer or mechanic . . . to work in surroundings or under working conditions which are unsanitary, hazardous, or dangerous to his health or safety, as determined under construction safety and health standards promulgated by the Secretary by regulation." Note that it will not be up to you to determine the conditions. On the one hand we have environmentalists driving for sanitation and clean-up, and on the other hand Congress is passing laws which prohibit anyone from performing any unsanitary work! It's like trying to make a mountain so big you can't move it; something here is going to turn out to be impossible.

No doubt there is justification for many of the standards and corrective action must be taken in many areas. It is in these areas that the contractor should concentrate his efforts, even without the legal necessity.

EFFECTS OF NOISE ON CONSTRUCTION WORKERS

The California State Department of Public Health published a very revealing and alarming report in 1966 on the deleterious effects of noise on the hearing of workers subjected to excessive noise. The time has come for contractors to take tangible steps toward the reduction of noise to safe levels. State and federal laws are being enacted rapidly since the problems were spotlighted by the passage of the Occupational Safety and Health Act of 1970.

The thing that is difficult to understand is that motorcycles without mufflers are allowed to run loose in residential areas, creating more noise than an entire construction project. Then on Sunday the construction worker is either out in a noisy motorboat, riding a noisy bike, or running a noisy lawnmower. When he goes deaf, the contractor is expected to foot the bill.

At last, however, Congress has passed anti-noise requirements limiting the noise levels on all gasoline motors and machines. But since this will not include those already manufactured, we'll all be deaf before the present ones wear out.

The law states that "in no case shall the prime contractor be relieved of the overall responsibility for compliance with the requirements . . . ." Therefore, he can not divert the responsibility to the subcontractor even for the sub's own actions. Accordingly, he will be responsible even for all equipment noise and other violations. In a study made by the California State Department of Public Health, 1966, only two out of the 25 pieces of heavy equipment studied were within the prescribed limits of noise level. Although equipment manufacturers are developing quieter equipment, something will have to be done about the noise in present equipment. Air conditioned and insulated cabs will help solve the problem on off-the-road construction machinery. Although this is a rather expensive conversion, it may pay off in increased productivity. After all, the reduction of dust, heat, noise, odors, and other distractions will certainly permit the operator to stay on the job longer and work more effectively.

CLEAN AIR REQUIREMENTS

Unfortunately, even the air conditioned cab is not the complete answer. These systems do not remove the ultra-low micron size particles from the air and these are the most damaging to the lungs. Particles in the 5-micron or less range must also be removed, either by chemical or other means. Respirators do only a partial job and, due to the discomfort to the wearer, are seldom used over any prolonged period of time. Serious lung damage can occur after only a few weeks' exposure to severe dust conditions. And, what is worse, the conditions are not always apparent. Dust in the 5-micron range is not visible in concentrations which can be serious or even fatal after prolonged exposure.

The problem is not solved by wetting down the soil, or by wet drilling or grinding, although this helps considerably. A small amount of wetting agent, such as ordinary detergent, added to the water materially aids in reducing airborne dust.

To further complicate the dust problem, some soils in arid areas contain spores which become airborne when the soil is disturbed. These spores cause a lung infection medically termed *coccidioidomycosis*. Susceptibility of workers may be determined by making skin tests. (Usually local residents build up a resistance to the disease.) Although seldom fatal, the disease can be incapacitating.

The law states, in effect, that "harmful dusts, fumes, vapors, or gases" are to "be controlled by the application of general ventilation, local exhaust ventilation, or other effective mechanical means." Furthermore, these are to be designed so that exhaust air is not "drawn through the work area of employees." To satisfactorily comply with this requirement could be very costly, and in many situations not at all practical. Many of the requirements have evidently been devised for manufacturing situations and need to be altered to fit construction operations.

FIRE PROTECTION

The employer is held responsible for the development of an adequate fire protection program, including equipment, training and facilities. Fire alarm devices and fire extinguishers must be properly located and rigid conditions for the storage of flammable materials must be met.

SIGNS, SIGNALS AND BARRICADES

It is not sufficient to simply mark or post the various signs and signals. Specific colors and designs for the various types of signs are stipulated. The law not only requires the signs and signals under specified conditions, but also requires that they be removed or covered as soon as the danger is past.

Now, get this one: "Signaling directions by flagmen shall conform to American National Standards Institute D6.1-1961, Manual on Uniform Traffic Control Devices for Streets and Highways." Red or orange garments must be worn by all flagmen.

If a tool or piece of equipment becomes defective, it must be immediately tagged with the proper warning tag to prevent its use by others, or the tool must be "physically removed from its place of operation." One big problem will be that of educating all employees and motivating them to comply with all of these rules.

When a workman gets jolted by an electrical short in a portable tool, he usually disconnects the tool and looks for a replacement. In the meantime, another worker grabs the tool for his own use. How are we going to be sure that the first worker will "properly, and immediately, tag the tool" before leaving it? Actually, the job of educating the workmen is the really big one.

MATERIALS HANDLING, STORAGE, USE, AND DISPOSAL

Maximum safe load limits of floors within buildings are required to be properly posted, and these limits are not to be exceeded in the placement of materials. This is not an easy one. How are the workmen to know the weights of lumber or other materials without weighing them? Compliance with a requirement such as this could be very time consuming, to say the least.

Noncompatible materials are to be segregated in storage. The question is: Which materials are noncompatible? Rules are set down for stacking brick, block, lumber, sacks of cement and other materials. To be in reasonable compliance, a contractor will have to employ at least one competent man full time, and even then, he will probably be the busiest man on the project if he accomplishes his job.

DISPOSAL OF WASTE MATERIALS

Whenever materials are dropped more than 20 feet to any point outside the walls of the building, an enclosed chute is to be used. Usually, it is not feasible to build such a chute from each floor; therefore, another way must be found to get waste materials down. It cannot be dropped through holes in the floor unless a barricade is set up around the area at least six feet larger all around than the opening in the floor above. Besides that, proper warning signs must be posted prior to dropping materials.

All solvent rags or other flammable materials are to be kept in covered containers until removed from the site.

HAND AND POWER TOOLS

Employers are not to issue or permit the use of unsafe tools. To prevent the workmen from using unsafe tools of his own, the employer would have to have a very knowledgeable safety man make a personal inspection of every workman's tools, and almost on a daily basis.

The use of electric cords for lowering or raising power tools is not to be permitted. Again, educating and changing the habits and practices of workmen will be the big job.

Compressed air is not to be used for cleaning purposes (with certain exceptions) unless reduced below 30 psi. In many instances, this renders the process ineffective. Even then, a chip guard and certain protective clothing must be used. All fixed powered woodworking tools are to be provided with switches which can either be locked or tagged in the off position.

ELECTRICAL REQUIREMENTS

Electrical cords and extensions on portable tools are to be of the three-wire grounded type (with certain exceptions). Extension cords are not to be hung from nails, fastened with staples, or suspended by wire. Also, extension lines are to be protected from pinching in doors, cuts by sharp corners or damage by traffic. I have yet to see my first job where these precautions have been taken. Portable lights operated in hazardous locations *must not be more than 12 volts.*

Splices in electrical cords *must be soldered.* "Cables passing through work areas shall be covered or elevated. . . ." Probably no construction job anywhere meets these requirements! Brass shell, paper-lined lamp sockets *are not to be used.* Special insulated tools must be used to change fuses if current in the box is live. No more changing fuses by hand or pliers.

LADDERS AND SCAFFOLDING

Portable ladders "shall be tied, blocked, or otherwise secured" to prevent their being displaced. It will take a little ingenuity to comply with this one in some instances. You are not allowed to build just any type of ladder. The law spells out the dimensions and contains many limitations. Not only that, a table of the various permissible species of woods is given and the specs vary depending upon the species of wood used. Question: Who is going to make positive identification of the wood? Even experts cannot identify many species without microscopic analysis and scientific classification. Douglas Fir (Rocky Mountain type) is in a different classification from Douglas Fir (Coast region), while White Fir is in still another classification! How many of your workmen can tell the difference?

It is not permissible to support scaffolds or planks on barrels, boxes, loose brick or block. Since no qualification is furnished as to height, it must be assumed that even a plank eight inches off the ground cannot be placed on blocks, boxes or other similar "unstable objects." To build a scaffold or ladder under these regulations, a skilled mechanic or engineer must refer to the various tables and limitations provided in the Bureau of Labor Standards. No longer can the carpenters simply go out and build scaffolds and ladders as they have in the past, because invariably they will not comply with the regulations.

CRANES, DERRICKS, HOISTS, ELEVATORS, AND CONVEYORS

The employer is required to designate a "competent person" who will inspect all machinery and equipment "prior to each use" and "during use" to make sure it is in safe operating condition. An annual inspection of hoisting machinery is required and *records must be kept*. Strict requirements are set down for the inspection of wire rope. Specific limits on the amount of wear are given and the table must be used when measuring for amount of wear on cables.

Exhaust pipes must be insulated or guarded where it is "possible for employees to make contact in normal duties." A fire extinguisher of the proper rating must be provided at the cab or operator station of all equipment.

Specified distance, depending upon the line voltage, is required when equipment is operated near high voltage lines. Also, a person shall be designated to observe the clearance of the equipment and "give timely warning" where operator visibility is restricted.

MOTOR VEHICLES AND MECHANIZED EQUIPMENT

Whenever equipment is parked the parking brake must be set. Equipment parked on inclines must have the parking brake set and the wheels chocked. This is a relaxation of the original rule that parking brakes and wheel chocks must be used every time the equipment is parked.

Tools and materials, when carried in the same compartment with employees, must be secured. This rule could become a real nuisance, for employees would not be allowed to carry a small tool kit in a vehicle without tying it down in some way.

Here's a good one. "All rubber tired vehicles shall be equipped with *fenders.*" Get busy, boys, and start making up fenders! A long checklist is given, and every piece of equipment is supposed to be checked over *at the beginning of each shift.*

This rule has since been relaxed somewhat and now states that "all rubber-tired motor vehicle equipment manufactured on or after May 1, 1972, shall be equipped with fenders." However, "all rubber-tired motor vehicle equipment manufactured before May 1, 1972, shall be equipped with fenders not later than May 1, 1973."

Another qualification is that "mud flaps may be used in lieu of fenders whenever motor vehicle equipment is not designed for fenders." It would seem to me that if the original equipment did not have fenders, it probably was not designed for fenders.

A recently added requirement is that "all bidirectional machines, such as rollers, compactors, front-end loaders, bulldozers, and similar equipment shall be equipped with a horn, distinguishable from the surrounding noise level, which shall be operated as needed when the machine is moving in either direction."

EXCAVATIONS, TRENCHING AND SHORING

Where employees are exposed to possible cave-ins, banks and trenches more than five feet (formerly four feet) in height or depth must be shored, laid back to a stable slope, or some other means of protection provided. Water must not be allowed to accumulate in excavations.

A table showing minimum requirements of trench shoring is provided in the regulations, but application could be a problem in some instances. For example, stringers are required where soil is "likely to crack." It would never be safe to assume that the soil was not likely to crack; therefore to stay within the law, stringers should *always* be used.

CONCRETE, CONCRETE FORMS AND REINFORCING

"Employees working more than six feet above any adjacent working surfaces placing and tying reinforcing steel in walls, piers, columns, etc., shall be provided with, and directed to wear safety belts, which are properly secured. . . ." Also, "Employees shall not be permitted to work above vertically protruding reinforcing steel unless it has been protected to eliminate the hazard of impalement."

When discharging on a slope, the wheels of concrete ready-mix trucks are to be blocked and the brakes set.

STEEL ERECTION

Rules and regulations state that "there shall not be more than eight stories between the erection floor and the uppermost permanent floor" (with certain qualifications).

On buildings not adaptable to temporary floors, and where scaffolds are not used, *safety nets shall be installed* whenever the distance exceeds two stories or 25 feet.

Where erection is being done by a crane on the ground, a tight and substantial floor shall be maintained within two stories or 25 feet, whichever is less, below and directly under that portion of each tier of beams on which bolting, riveting, welding or painting is being done.

"When bolts or drift pins are being knocked out, means shall be provided to keep them from falling."

Planking or decking of equivalent strength shall be of proper thickness to carry the working load. "Planking shall not be less than two inches thick full size undressed. . . ." This could present a problem, and may require going to the next larger thickness to get the full two-inch measurement.

Let me reiterate, this digest does not provide complete coverage of the law. The complete law consists of *several thousand regulations*.

TUNNELS AND SHAFTS

The regulations state that "access to unattended underground openings shall be restricted by gates or doors. Unused chutes, manways, or other openings shall be tightly covered, bulkheaded or fenced off, and posted."

"Each operation shall have a check-in and check-out system that will provide positive identification of every employee underground. An accurate record and location of the employees shall be kept on the surface." These rules are much the same as for mines, and are very stringent.

"Telephone or other signal communication shall be provided between the work face and the tunnel portal, and such systems shall be independent of the tunnel power supply."

The law is quite specific regarding quality of air and oxygen content. The air must be analyzed frequently and tested for various gases and contaminents. "Respirators shall not be substituted for environmental control measures." Regarding ventilation, "Tunnels shall be provided with mechanically induced primary ventilation in all work areas." The law further states that "the direction of airflow shall be reversible."

"Ventilation doors, not operated mechanically, shall be designed and installed so that they are self-closing and will remain closed regardless of the direction of the air movement."

As can be seen from these brief and partial excerpts, the rules and regulations are detailed and exacting. A qualified consultant, or trained safety engineer, would probably have to be employed to see that these requirements are fulfilled.

COMPRESSED AIR

The law states that "there shall be present, at all times, at least one competent person designated by and representing the employer, who shall be familiar with this subpart [regulations on the use of compressed air] in all respects, and responsible for full compliance with these and other applicable subparts." Also "every employee shall be instructed in the rules and regulations which concern his safety or the safety of others."

DEMOLITION

Many of the requirements for demolition are the same as those for building construction. In addition to the ordinary precautions, an engineering survey is required to determine whether danger exists of collapse of walls, floors, or any other part of the structure. Also, it must be determined whether any harmful chemicals, gases, or other hazards are present.

All utility services must be disconnected, including *sewer*. The rules state, "In any case, any utility company which is involved shall be notified in advance. . . . Only those stairways, passageways, and ladders designated as means of access to the structure of building, shall be used. Other access ways shall be entirely closed off at all times." Also, "All material chutes . . . at an angle of more than 45° from horizontal, shall be entirely enclosed, except for openings equipped with closures at or about floor level for the insertion of materials. The openings shall not exceed 48 inches in height. . . ." At all stories below the top floor, openings must be kept closed when not in use. The extra time involved here is very obvious.

Requirements are stipulated for removal of walls, floors, chimneys and other masonry. Workmen are not allowed to walk on beams. Special plank walks are to be built and used. Where a "headache ball" is used in demolition, the ball must not exceed 50 percent of the crane's rated load, based on the length of the boom and the maximum angle of operation, or it is not to exceed 25 percent of the breaking strength of the line by which it is suspended, whichever is less.

All roof cornices or other ornamental stonework must be removed before walls are pulled over.

BLASTING AND THE USE OF EXPLOSIVES

Obviously, no person is to handle explosives other than one fully qualified. Even a qualified person is not to handle explosives while under the influence of liquor, drugs, or medicines which affect mental capabilities.

No fire is to be fought where the fire is in imminent danger of contact with explosives, and all personnel are to be removed to a safe distance.

Precautions are to be taken to prevent the accidental discharge of blasting caps from current induced by radar, radio transmitters, lightning, power lines, dust storms or other sources of extraneous electricity. Warning signs are to be posted within 1,000 feet of the blasting operations. All mobile radio transmitters within 100 feet of blasting caps *are to be de-energized and locked.* The use of black powder *is prohibited.*

SOURCES OF ASSISTANCE ON SAFETY AND HEALTH

Again, this treatise is not intended to be a complete guide on the Occupational Safety and Health Law, but is intended to convey some of the important facets of this complex law and to stress the seriousness of the costs involved in compliance. A competent safety engineer will usually be required to assure full compliance since the law contains thousands of regulations. The details of the law would require weeks or months of study even for one previously trained in this work.

EFFECTIVE DATES

The law is effective and applies to contracts on and after April 24, 1971, and to all projects negotiated on or after April 27, 1971. Some of the regulations dealing with light residential construction became effective September 27, 1971. "Light residential construction" means homes and apartments not over three stories high, which do not have an elevator.

PEOPLE CREATE THEIR OWN HAZARDS

Machinery doesn't maim and kill; people do. Manufacturers cannot design and build machines which are completely safe. The operator, the maintenance man, and others create their own hazards. Safety is a dull subject. It's like insurance—nobody gives it a thought until it is too late. But, if you want to develop an efficient organization, if you want to keep costs under control, if you want to maintain a tight schedule, then the time you spend developing a good safety program will be an excellent investment. And, best of all, you won't have to lie awake nights wondering what you should have done—or *could* have done—to prevent that serious accident.

There are many costly side effects to injuries or death on the job. There is always bad publicity resulting from accidents or failures of any type. Morale of the entire work force hits a serious low. Productivity always suffers, not to mention the suffering of the injured workers and their families.

Remember that construction operations constitute the greatest hazard of all types of industries—all the more reason why safety efforts should be increased.

Now exactly *what* can be done in addition to taking normal safety precautions? We assume you have taken such precautions with the operation of equipment. You have probably posted safety posters in prominent places, and sent safety messages around pertaining to specific hazards. What else can be done? Accidents continue to happen. Why? Carelessness? Sometimes.

The foreman or supervisor on the job must be constantly safety conscious. I can visit almost any project and point out safety hazards—accidents waiting to happen. If they are so evident, the foreman or supervisor certainly can see them too! I see lumber lying around with nails sticking straight up. My dad would have skinned me alive if I had thrown a board down with a nail sticking up. Not only that, he would have been just as rough on me if I had even *passed* one without correcting the condition. There are certain things you just *don't do;* and most of us know what they are. Obviously, I can't list all of them here. Don't take unnecessary risks and don't let the men do it.

Recently one of the men operating a table saw was seen running material through the saw *backwards!* I noticed this from a considerable distance away and could hardly believe my eyes. I asked him why he did this and he said, "It makes a cleaner cut. The boss said the cuts were too splintery."

This is an instance where the "boss" should have followed through with more complete communication. Actually, what the operator needed was a saw blade in good condition. He didn't have a better blade and was trying to get the job done the best way he could. Further examination of the situation revealed that the machine was in poor condition. The guide was not in alignment with the saw blade. After getting the machine in good condition, there was no reason for makeshift, improvised, dangerous operation.

The foreman should have determined the source of the problem and given specific advice concerning the remedy. The rule is: *Never toss a problem at a workman which you cannot answer yourself*. See, "How to Develop Your Problem-Solving Capability" in Chapter 11.

THE HAZARDS OF USING UNSAFE EQUIPMENT

While I was writing this chapter on safety, a man came out to install a soft water system. After showing him where to put it, I went back to work. In a few minutes he knocked on the door and asked if he could connect his electric drill into a receptacle in the entryway, saying, "There's no current in the outlet out here." I plugged his line into the inside receptacle, but there was still no life from the drill. In the first place, the line was a two-conductor with no ground; it was also too light for the length of line, its insulation was worn off, and bare wires showed through in places.

While the man was out at his truck, I removed the extension line and plugged the drill into the outside receptacle. Still no life. I flipped the reversing switch and sparks flew, but it did make contact. I could count at least four violations of safety rules in that one operation. I wonder how many I could count if I followed him around all day!

Next, I telephoned the worker's manager and described the situation. He said, "The boys are supposed to turn in any bad equipment for repair." But the equipment furnished wasn't even the proper equipment in the first place!

Workmen usually try to use the equipment furnished them. They know that if they complain about every problem, they will become unpopular. So, as I pointed out to the manager, it is management's responsibility to furnish safe and proper equipment in the first place. Then, educate the workmen to be careful with the equipment and keep it in safe operating condition. Next make a periodic inspection of all equipment. Don't expect the workmen to report on the condition of equipment. Usually, they won't report it until *total failure is experienced*. In the meantime, they are flirting with serious hazards.

SOME UNDERLYING CAUSES OF ACCIDENTS

Don't forget that poor eyesight is the cause of many accidents. Other causes of accidents are poor hearing, slow reaction time and lack of coordination. Many medicines cause sluggishness in motor reflexes. Keep these things in mind when analyzing the causes of accidents. Close observation of workers may reveal some of the above handicaps in time to prevent accidents. Emotional pressure is one of the main causes of accidents; it makes the individual accident-prone because of his inability to concentrate on his work. When his mind is not on his work, his defense mechanism is impaired.

Here is my recommendation: Set aside one day which we will call "Safety Day." Concentrate all efforts on the workers' activities, analyzing them to locate those individuals who seem to be disorganized or who show evidence of a lack of coordination. You don't have to "put them on the couch." Just observe without staring. All experienced managers can "see without looking." They used to tell me I had eyes in the back of my head. Look for signs of danger such as loose scaffolding or walk boards, unstable piles of material, makeshift repairs to machines and equipment, loose parts on equipment, debris on the floors of work areas (such as the saw shed), congested isles, protruding nails, unprotected excavations, lack of proper safety clothing where needed, etc.

Then call a brief meeting at the end of the day and discuss the results of your day's observation. Tell the workers what you have found and let them know that their own safety is being impaired by the violations you discovered. Let them know that each one of them must become a safety committee of one. The carelessness of one man can endanger many.

If the accident rate drops, you know your efforts have been rewarded. Then don't relax your efforts, but plan a regular program. Of course, safety is a constant endeavor, but a weekly program such as the one I have suggested will probably make a noticeable improvement over previous records.

PSYCHOLOGICAL FACTORS AFFECTING SAFETY

Construction management should always be on the alert for the accident-prone individual. He is the man who always seems to have his mind on something other than his work. He performs his work in a haphazard manner, and often fails to get his instructions correct. This man should be removed from hazardous operations for his own safety and that of others as well. If the condition is temporary, obviously the technique of handling the situation will be different. A private talk with the employee may bring his problems to the surface. There may be something you can do to help alleviate his problem. Showing your interest in him will usually help. The dangerous aspects of his unstable condition should be stressed so that he will make a special effort to concentrate on his work.

ACCIDENT REPORTS

At the end of this chapter you will find four different forms for reporting on various types of vehicle accidents. Since they are self-explanatory, no discussion is necessary. For more about forms and form design see Figures 10-4, 10-5, 10-6 and 10-7.

KEEPING UP TO DATE ON THE LAW

It is imperative that those firms who are affected by the Occupational Safety and Health law keep current on the numerous and constant changes taking place in the law. Prentice-Hall, Inc., Englewood Cliffs, N.J., provides a labor relations service with a loose-leaf revision service to keep it current. This, and other sources of information, are described in Chapter 12.

STANDARD FORM 91-A (REV. JUNE 1953; ED. 1955)

INVESTIGATION REPORT OF MOTOR VEHICLE ACCIDENT

(Department or establishment)

1.

(Name and location of reporting unit)

2. GENERAL LOCATION, DATE, DAY AND HOUR OF ACCIDENT

IF ACCIDENT IN CITY, GIVE CITY OR TOWN AND STATE; IF OUTSIDE CITY LIMITS, INDICATE MILEAGE OR DISTANCE TO NEAREST CITY OR TOWN

(City or town) — (Miles) (Direction) FROM: Limits ☐ Center ☐

(County and State)

DATE	DAY OF WEEK	HOUR A. M. P. M.

3. EXACT LOCATION OF ACCIDENT

ACCIDENT OCCURRED ON (Street) (Highway)

NOTE: CHECK AND COMPLETE ONE. Name (or otherwise identify) nearest intersecting street, house number, power or telephone pole (give number), highway curve, bridge, railroad crossing, filling station, alley, driveway, culvert, guardrail, milepost, underpass, or other identifying landmark. Show exact distance.

☐ AT INTERSECTION WITH (Street or alley)

☐ NOT AT INTERSECTION (Distance) (Direction) of

AND (Distance) (Direction) of

4. FEDERAL VEHICLE (Fed.) (Includes Privately Owned Federally Operated)			5. OTHER VEHICLE (2)		
YEAR	MAKE	BODY TYPE	YEAR	MAKE	BODY TYPE
REGISTRATION NO.	KIND OF CARGO	NUMBER OF PASSENGERS	REGISTRATION NO.	KIND OF CARGO	NUMBER OF PASSENGERS
WAS CARGO DAMAGED? YES		NO	WAS CARGO DAMAGED? YES		NO
PARTS OF VEHICLE DAMAGED AND NATURE OF DAMAGE			PARTS OF VEHICLE DAMAGED AND NATURE OF DAMAGE		
GOING (Direction)	ON (Street or highway)		GOING (Direction)	ON (Street or highway)	
DISTANCE DANGER NOTICED (Feet)	ESTIMATED SPEED THEN (m. p. h.)	ESTIMATED SPEED AT IMPACT (m. p. h.)	DISTANCE DANGER NOTICED (Feet)	ESTIMATED SPEED THEN (m. p. h.)	ESTIMATED SPEED AT IMPACT (m. p. h.)
LAWFUL SPEED (m. p. h.)	DISTANCE TRAVELED AFTER IMPACT (Feet)		LAWFUL SPEED (m. p. h.)	DISTANCE TRAVELED AFTER IMPACT (Feet)	
MAXIMUM SAFE SPEED (m. p. h.)	OPERATOR'S PERMIT ☐ FEDERAL	☐ STATE	MAXIMUM SAFE SPEED (m. p. h.)	OPERATOR'S PERMIT	
TYPE OF PERMIT (Issuing State) (Permit number) ☐ CHAUFFEUR	☐ TRUCK DRIVER	☐ OPERATOR	TYPE OF PERMIT (Issuing State) (Permit number) ☐ CHAUFFEUR	☐ TRUCK DRIVER	☐ OPERATOR
LIMITATION OF PERMIT			LIMITATION OF PERMIT		
DRIVER'S NAME		SEX / AGE	DRIVER'S NAME		SEX / AGE
ADDRESS			ADDRESS		
NUMBER OF HOURS ON DUTY PRECEDING ACCIDENT	YEARS DRIVING EXPERIENCE	EXPERIENCE THIS TYPE VEHICLE	NAME AND ADDRESS OF OWNER (Include phone number)		

BUREAU OF THE BUDGET CIRCULAR A-5 (REV.) 91—204

6. WITNESSES

A. NAME	PHONE NO.	B. NAME	PHONE NO.
ADDRESS		ADDRESS	
LOCATION OF WITNESS AT TIME OF ACCIDENT		LOCATION OF WITNESS AT TIME OF ACCIDENT	

7. KILLED OR INJURED

A. NAME	SEX / AGE	B. NAME	SEX / AGE
ADDRESS		ADDRESS	
CHECK ONE: ☐ KILLED ☐ INJURED — CHECK ONE: ☐ DRIVER ☐ PASSENGER ☐ HELPER ☐ PEDESTRIAN		CHECK ONE: ☐ KILLED ☐ INJURED — CHECK ONE: ☐ DRIVER ☐ PASSENGER ☐ HELPER ☐ PEDESTRIAN	
CHECK ONE ☐ IN FED VEHICLE ☐ IN OTHER VEHICLE — WHERE IN VEHICLE?		CHECK ONE ☐ IN FED. VEHICLE ☐ IN OTHER VEHICLE — WHERE IN VEHICLE?	
FIRST AID GIVEN BY		FIRST AID GIVEN BY	
TAKEN TO		TAKEN TO	
TAKEN BY	REGISTRATION NO.	TAKEN BY	REGISTRATION NO.

8. PEDESTRIAN

PEDESTRIAN WAS GOING ☐ ON ☐ ACROSS (Direction) (Street, highway No., etc.) FROM (SW. cor., W. side, etc.) TO (NE. cor., W. side, etc.)

PEDESTRIAN WAS (Check one)

- ☐ 1. CROSSING AT INTERSECTION WITH SIGNAL
- ☐ 2. SAME—AGAINST SIGNAL
- ☐ 3. SAME—NO SIGNAL
- ☐ 4. SAME—DIAGONALLY
- ☐ 5. CROSSING NOT AT INTERSECTION COMING FROM BEHIND PARKED CARS
- ☐ 6. SAME—NOT COMING FROM BEHIND PARKED CARS
- ☐ 7. COMING FROM BEHIND PARKED CARS TO ENTER VEHICLE
- ☐ 8. WAITING OR GETTING ON OR OFF AT STREET CAR SAFETY ZONE
- ☐ 9. NOT AT SAFETY ZONE
- ☐ 10. GETTING ON OR OFF ANOTHER VEHICLE
- ☐ 11. PLAYING IN ROADWAY
- ☐ 12. WORKING IN ROADWAY
- ☐ 13. WALKING IN ROADWAY—WITH TRAFFIC
- ☐ 14. WALKING IN ROADWAY—AGAINST TRAFFIC
- ☐ 15. WALKING IN ROADWAY—SIDEWALKS AVAILABLE
- ☐ 16. WALKING IN ROADWAY—NO SIDEWALKS AVAILABLE
- ☐ 17. HITCHING ON VEHICLE
- ☐ 18. LYING IN ROADWAY
- ☐ 19. NOT IN ROADWAY (Explain)

9. DAMAGE TO PROPERTY OTHER THAN MOTOR VEHICLES OR CARGO

NAME OBJECTS, SHOW OWNERSHIP, STATE NATURE OF DAMAGE

10. KIND OF LOCALITY (Check one)		11. LIGHT (Check one)	12. WEATHER (Check one)	
☐ 1. MANUFACTURING AND INDUSTRIAL	☐ 5. OPEN COUNTRY	☐ 1. DAYLIGHT ☐ 3. DAWN	☐ 1. CLEAR	☐ 4. FOG
☐ 2. SHOPPING AND BUSINESS	☐ 6. INDUSTRIAL PREMISES	☐ 2. DUSK	☐ 2. RAINING	☐ 5. OTHER (Specify)
☐ 3. RESIDENTIAL	☐ 7. HOME OR DOMESTIC PREMISES	DARKNESS WITH:	☐ 3. SNOWING	
☐ 4. SCHOOL AND PLAYGROUND	☐ 8. OTHER (Specify)	☐ 4. ARTIFICIAL LIGHT		
		☐ 5. NO ARTIFICIAL LIGHT		

FIGURE 10-4

Investigation Report of Motor Vehicle Accident

FIGURE 10-4 (cont.)

OPERATOR'S REPORT OF MOTOR-VEHICLE ACCIDENT

THIS FORM TO BE FILLED OUT BY THE GOVERNMENT OPERATOR AT THE TIME AND AT THE SCENE OF THE ACCIDENT, INSOFAR AS POSSIBLE.

DEPARTMENT OR AGENCY

NAME AND LOCATION OF ORGANIZATION TO WHICH YOU ARE ASSIGNED

I OPERATOR

PLEASE PRINT FULL NAME	RANK, RATING, OR TITLE
SERVICE NUMBER OR ROSTER TITLE	OPERATOR'S STATE PERMIT NUMBER
HOME ADDRESS (*Street, city, State*)	TELEPHONE (*Home*)

II TIME AND PLACE

DATE AND DAY OF WEEK OF ACCIDENT	HOUR (*A.M. or P.M.*)

PLACE OF ACCIDENT (*If accident in city, give city and State; if outside city limits, indicate mileage to nearest city or other landmark.*)

FROM WHAT PLACE TO WHAT PLACE WERE YOU BOUND

FOR WHAT PURPOSE

III YOUR VEHICLE

MAKE	TYPE	REGISTRATION NUMBER OR OTHER IDENTIFICATION

PARTS OF VEHICLE DAMAGED (*Describe*)

ESTIMATED AMOUNT OF DAMAGE $

IV OTHER VEHICLE AND PROPERTY (For additional vehicles see page 3)

MAKE	TYPE	YEAR
OPERATOR'S STATE PERMIT NUMBER	VEHICLE LICENSE NUMBER	
OPERATED BY (*Name*)	VEHICLE OWNED BY	
OPERATOR'S HOME ADDRESS (*Street, city, State*)	OWNER'S ADDRESS (*Street, city, State*)	

PARTS OF VEHICLE DAMAGED (*Describe*)

ESTIMATED AMOUNT OF DAMAGE $

OTHER VEHICLE OR PROPERTY DAMAGED (*Describe*)

XIII OPERATOR'S STATEMENT OF ACCIDENT AND USE OF SAFETY EQUIPMENT

TELL IN YOUR OWN WAY HOW ACCIDENT HAPPENED

WAS VEHICLE EQUIPPED WITH SEAT BELTS? ☐ YES ☐ NO

IF YES, WERE THEY IN USE AT TIME OF ACCIDENT? ☐ YES ☐ NO

SIGNATURE OF OPERATOR	DATE

HAVE YOU ANSWERED ALL QUESTIONS AS COMPLETELY AS POSSIBLE?

Standard Form 91
Revised April 1963

(91–105)

BUREAU OF THE BUDGET
Circular A-5 (Rev.)

☆ U.S. GOVERNMENT PRINTING OFFICE : 1963 O—706-356

4

FIGURE 10-5

Operator's Report of Motor Vehicle Accident

FIGURE 10-5 (cont.)

COMPLYING WITH THE OCCUPATIONAL HEALTH AND SAFETY REQUIREMENTS

Standard Form 95
Revised February 1963
Bureau of the Budget
Circular A-5 (Rev.)

SUBMIT TO:

CLAIM FOR DAMAGE OR INJURY
(Use additional sheets if necessary)

95–104

Use ink or typewriter. See reverse side for instructions and additional information required.

1. NAME OF CLAIMANT (Please print full name)	2. AGE	3. MARITAL STATUS	8. AMOUNT OF CLAIM	
4. ADDRESS OF CLAIMANT (Street, city, zone, State)			PROPERTY DAMAGE	$
5. NAME AND ADDRESS OF SPOUSE, IF ANY			PERSONAL INJURY	$
6. PLACE OF ACCIDENT (Give city or town and State; if outside city limits, indicate mileage or distance to nearest city or town)			TOTAL	
7. DATE AND DAY OF ACCIDENT	TIME (A.M. or P.M.)			$

9. DESCRIPTION OF ACCIDENT—STATE BELOW, IN DETAIL, ALL KNOWN FACTS AND CIRCUMSTANCES ATTENDING THE DAMAGE OR INJURY, INDENTIFYING PERSONS AND PROPERTY INVOLVED AND THE CAUSE THEREOF

PROPERTY DAMAGE

10.
| NAME OF OWNER, IF OTHER THAN CLAIMANT | ADDRESS OF OWNER, IF OTHER THAN CLAIMANT |

BRIEFLY DESCRIBE KIND AND LOCATION OF PROPERTY AND NATURE AND EXTENT OF DAMAGE. SEE INSTRUCTIONS ON REVERSE SIDE FOR METHOD OF SUBSTANTIATING CLAIM

PERSONAL INJURY

11.
STATE NATURE AND EXTENT OF INJURY WHICH FORMS THE BASIS OF THIS CLAIM

WITNESSES

12.
| NAMES | ADDRESSES |

CRIMINAL PENALTY FOR PRESENTING FRAUDU-
LENT CLAIM OR MAKING FALSE STATEMENTS

Fine of not more than $10,000 or imprisonment for not more than 5 years or both. (See 62 Stat. 698, 749; 18 U.S.C. 287, 1001.)

CIVIL PENALTY FOR PRESENTING
FRAUDULENT CLAIM

The claimant shall forfeit and pay to the United States the sum of $2,000, plus double the amount of damages sustained by the United States. (See R.S. §3490, 5438; 31 U.S.C 231.)

13. I DECLARE UNDER THE PENALTIES OF PERJURY THAT THE AMOUNT OF THIS CLAIM COVERS ONLY DAMAGES AND INJURIES CAUSED BY THE ACCIDENT ABOVE DESCRIBED. I AGREE TO ACCEPT SAID AMOUNT IN FULL SATISFACTION AND FINAL SETTLEMENT OF THIS CLAIM.

SIGNATURE OF CLAIMANT

DATE OF CLAIM

NOTE: Signature used above should be used in all future correspondence.

FIGURE 10-6

Claim for Damage or Injury

NOTICE TO CLAIMANT

In order that your claim for damages may receive proper consideration you are requested to supply the information called for on both sides of this form. All material facts should be stated on this form, as it will be the basis of further action upon your claim. The instructions set forth below should be read carefully before the form is prepared.

INSTRUCTIONS

Claims for damage to or for loss or destruction of property, or for personal injury, must be signed by the owner of the property damaged or lost or the injured person. If, by reason of death, other disability or for reasons deemed satisfactory by the Government, the foregoing requirement cannot be fulfilled, the claim may be filed by a duly authorized agent or other legal representative, provided evidence satisfactory to the Government is submitted with said claim establishing authority to act.

If claimant intends to file claim for both personal injury and property damage, claim for both must be shown in item 8 on this form. Separate claims for personal injury and property damage are not acceptable.

The amount claimed should be substantiated by competent evidence as follows:

(*a*) In support of claim for personal injury or death, the claimant should submit a written report by the attending physician, showing the nature and extent of injury, the nature and extent of treatment, the degree of permanent disability, if any, the prognosis, and the period of hospitalization, or incapacitation, attaching itemized bills for medical, hospital, or burial expenses actually incurred.

(*b*) In support of claims for damage to property which has been or can be economically repaired, the claimant should submit at least two itemized signed statements or estimates by reliable, disinterested concerns, or, if payment has been made, the itemized signed receipts evidencing payment.

(*c*) In support of claims for damage to property which is not economically reparable, or if the property is lost or destroyed, the claimant should submit statements as to the original cost of the property, the date of purchase, and the value of the property, both before and after the accident. Such statements should be by disinterested competent persons, preferably reputable dealers or officials familiar with the type of property damaged, or by two or more competitive bidders, and should be certified as being just and correct.

Any further instructions or information necessary in the preparation of your claim will be furnished, upon request, by the office indicated at the top of the other side of this form.

INSTRUCTIONS REGARDING INSURANCE COVERAGE

In order that subrogation claims may be adjudicated, it is essential that the claimant provide the following information regarding the insurance coverage of his vehicle:

DO YOU CARRY COLLISION INSURANCE? ☐ YES ☐ NO	IF YES, GIVE NAME AND ADDRESS OF INSURANCE COMPANY AND POLICY NUMBER
HAVE YOU FILED CLAIM ON YOUR INSURANCE CARRIER IN THIS INSTANCE, AND IF SO, IS IT FULL COVERAGE OR DEDUCTIBLE?	IF DEDUCTIBLE, STATE AMOUNT

IF SUCH CLAIM HAS BEEN FILED, WHAT ACTION HAS YOUR INSURER TAKEN, OR WHAT ACTION DOES IT PROPOSE TO TAKE WITH REFERENCE TO YOUR CLAIM? (*It is necessary that you ascertain these facts*)

DO YOU CARRY PUBLIC LIABILITY AND PROPERTY DAMAGE COVERAGE? ☐ YES ☐ NO	IF YES, GIVE NAME OF INSURANCE CARRIER

SIGNATURE OF CLAIMANT

U.S. GOVERNMENT PRINTING OFFICE : 1963—O-653283 #43-C

FIGURE 10-6 (cont.)

Standard Form 92-A
PROMULGATED JUNE 1953 BY BUREAU OF THE BUDGET
CIRCULAR A-5 (REV.) 92-203

REPORT OF ACCIDENT OTHER THAN MOTOR VEHICLE

DEPARTMENT OR AGENCY	ACCIDENT OCCURRED IN—	
NAME AND ADDRESS OF LOCAL REPORTING ORGANIZATION	GOVERNMENT OPERATION	CONTRACTOR OPERATION

TIME AND PLACE OF ACCIDENT	DATE OF ACCIDENT	TIME	EXACT LOCATION OF ACCIDENT

PERSON OR PERSONS INJURED

NAMES	ADDRESSES

NATURE AND EXTENT OF INJURIES

PROPERTY DAMAGED

NAME OF OWNER	ADDRESS OF OWNER

DESCRIPTION AND LOCATION OF PROPERTY

NATURE AND EXTENT OF DAMAGE

FULL DESCRIPTION OF ACCIDENT *(Use Reverse Side, If Necessary)*

WITNESSES

IMPORTANT.—Be particular to secure the names and addresses of witnesses, bystanders, or persons in the immediate vicinity who may have seen the accident or heard any statement made by the person injured.

NAMES	ADDRESSES

DATE OF THIS REPORT	TITLE OR POSITION	SIGNATURE OF PERSON EXECUTING THIS FORM

★ U.S. GOVERNMENT PRINTING OFFICE : 1969 O - 366-997

FIGURE 10-7

Report of Accident Other Than Motor Vehicle

11

How Other Contractors Have Solved Tough Technical Problems

In the construction business, management people should train themselves to become problem solvers, or look for a simpler occupation. You could attend some of the special classes at Harvard University and learn problem-solving techniques, but you don't have to. You already have plenty of problems. Don't ignore them. Dive into them one at a time and develop solutions. You will find that the solutions will come more and more easily as you gain experience in problem-solving. And there are plenty of other rewards, too, like lower operating costs, better safety record, improved scheduling, higher employee morale—all resulting from fewer frustrations and the satisfaction of turning out a better job.

HOW TO DEVELOP YOUR PROBLEM-SOLVING CAPABILITY

A person can develop his problem-solving capability simply *by solving problems*. He does not become a good problem solver by shunting the problems off to someone else. By nature, I am a determined and inquisitive individual. I have a natural curiosity which compels me to solve problems whenever and wherever I encountered them. I'll admit that at times this becomes rather taxing, but the constant practice of problem-solving has enabled me to quickly recognize and anticipate problems. Then, by a process of mentally reviewing numerous *possible* approaches, I invariably come up with a good solution. I bring this out simply to show that the mental gymnastics of problem-solving is a healthy way to develop mental capacity for solving complex problems.

The first step in problem solving is to get the facts—*all of the facts*. Otherwise, you may jump to an erroneous conclusion which may be worse than no solution at all.

Study everything associated with the problem. Most people attack the *symptoms* instead of the *problem*. And before you look for the solution, stop and ask yourself whether *you* are part of the problem. Have you done or said anything that could have influenced the workman's technique or method? When all of the facts are brought to light, the solution is usually self-evident. Break the problem down into its smallest components and analyze each part carefully. If the solution is not evident, back up and try a completely new approach. Problem-solving is essentially the same everywhere whether it is in construction or in the automobile assembly plant. The basic requirements for a problem solver are:

> Imagination
> Ingenuity
> Determination
> Persistence

These four ingredients, effectively applied, can solve almost any problem!

The very nature of the construction business brings on a never-ending procession of problems. Even two identical projects built in different locations are built under different conditions and may have entirely different problems. For many years I have assisted the construction industries in solutions to their problems, and I have developed solutions to hundreds of tough problems which were believed by many to have no solution.

The following problems and solutions are typical of the hundreds encountered on building and construction projects in recent years. You have probably encountered several of these and will probably encounter others. To facilitate the location of certain problems, they are indexed in this book under the key words. For example: problems of cracks in terrazzo floors are listed under "floors" and also under "terrazzo," as well as under "cracks."

PROBLEM: How can we lower a heavy piece of equipment into a pit without the use of a large crane?

SOLUTION: Fill the pit with 100-pound cakes of ice. Then move the equipment over on top of the ice and wait for the ice to melt. If you wish to accelerate the melting, the ice can be salted. Some precautions are in order here to prevent uneven melting and possible tipping of the equipment to be lowered.

Is the equipment mass fairly equally distributed? If the weight on the ice is much heavier on one side or one end, that side will melt faster and cause a possible tipping condition. To overcome this, salt may be concentrated on the lighter side to accelerate the melting on that side.

Rate of lowering will depend upon several factors:

> Height of ice to be melted.
> Weight of the object to be lowered.
> Ambient temperature.

Caution: If the melting water can damage the equipment, a small sump pump may be used to pump out water as it melts. This would be a good idea anyway if the pit is bare earth. Remember also that salt has a corrosive effect on metals. Wiping with oil will aid in protecting the equipment.

PROBLEM: Specifications call for tempered glass on a project we are completing. The glass has been installed, but how can we determine if it is tempered?

SOLUTION: A sheet of tempered glass can be identified by the tong marks on one edge. If these marks have been concealed during installation, the glass can be tested with a rubber mallet. Give the glass a hard blow with the rubber mallet. The strength of tempered glass is high enough to make it almost impossible to break with a rubber mallet. Be prepared, however, in the event the glass has not been tempered. Wear goggles and protect the hand and arm against glass fragments, since untempered glass will break under a heavy blow with a rubber mallet.

Untempered glass is easily scored with a glass cutter, whereas tempered glass offers high resistance to scoring. This is not recommended as a test on glass which has already been installed, since any score mark on tempered glass will probably cause the entire glass to shatter.

Glass is tempered by suspending it on tongs clamped to the top edge and lowering it into a furnace where it is heated almost to the melting point. It is withdrawn and cooled suddenly by a blast of air. The sudden cooling sets up high stresses in the outer surfaces. Tempered glass cannot be cut after tempering, therefore it must be made up in the exact size to be used. If the surface is damaged in any way, the internal stresses cause the entire sheet to fracture into small fragments.

PROBLEM: How can we control water seepage through masonry walls and floors below grade?

SOLUTION: First, be sure that moisture accumulation is not caused by condensation as is frequently the case in areas of high humidity. The solution here would be to install dehumidifying equipment or air conditioning. Remember that if the concrete has been recently poured, much of the moisture is from curing of the concrete; therefore, the condition may be temporary. If it is positively determined that moisture is leaking through the walls or floor, these may be sealed from the inside without the necessity of digging all around the outside of the basement walls to apply waterproofing. If seepage is experienced through the floor, sealing the walls outside will not solve the problem.

Leaks may be sealed from inside with one of the many epoxies now available. Surfaces must be absolutely dry before applying the epoxy and during the cure. Epoxies are available which will set in short or long periods of time, depending upon the requirements. Many alloys have been compounded to fill specific needs. To locate the best source in your area, just look in the index of the classified telephone directory. Several classifications are shown in most directories under the heading "Epoxy Adhesives and Cements." These are listed in many of the larger directories under "Concrete Patching Compounds" and "Waterproofing Materials."

Be sure surfaces are dry and clean, and that all loose particles have been removed. Then just follow the instructions given with the material. A true epoxy is a two-component material to be mixed just prior to application. Epoxy esters are single component materials and are not true epoxies.

PROBLEM: We are experiencing a problem with cracks and open joints in the interior woodwork in a building completed last November. The supplier of the woodwork claims that moisture content of the materials used was between 8 and 10 percent when delivered to the job. The building was not occupied until late December due to delays caused by plaster not drying and heating system not in operation.

SOLUTION: The problem arises in that the woodwork has absorbed moisture after being delivered to the job. Even though the woodwork may have been primed, it can still absorb moisture under conditions of high humidity. When the wood (tightly fitted in place) absorbs moisture it swells and compresses the fiber of the wood, resulting in some crushing of the fiber. Once the fiber is crushed, it does not return to the original size, thereby leaving open cracks and joints after drying. This is technically known as *compression set*.

More information on this problem and the proper handling and storage of interior woodwork items will be found in Chapter 8.

PROBLEM: What can we do to prevent dusting of concrete floors?

SOLUTION: Dusting of concrete is due to weakness in the concrete at the surface. This weakness is usually due to too-wet concrete mix and troweling of bleed water on the surface. Inadequate curing can also cause dusting problems. Even with a good mix (not over 4-inch slump), overtroweling or troweling at the wrong time can cause dusting. Excessive troweling, particularly while concrete has bleed water on the surface, works water and fines to the surface and leaves a weak, diluted mortar on the surface.

On the other hand, inadequate curing, allowing too rapid evaporation of the surface moisture, halts the chemical process and leaves the surface weak. The slightest abrasion of such a surface loosens this weak top layer, converting it into dust.

The following procedure is recommended to avoid this problem:

1. Use a relatively dry mix with not over 4-inch slump. Water content should not exceed 5½ gallons per sack of cement.
2. Concrete should be placed on a dampened subgrade to prevent loss of water to the subgrade. It should not be flowed into place; instead, it should be chuted or wheeled into place.
3. After the concrete has been leveled and tamped in place, a strikeoff screed may be used. The surface may be dressed with a float after the water sheen has disappeared. (For air-entrained concrete, use a float 30 to 40 minutes after placing.) *Never do any finishing while bleed water is showing!* Final troweling should begin when consistency

of concrete is such that finger pressure just dents the surface. A steel trowel will then produce a ringing sound.

4. Surface of concrete must be kept moist to allow normal chemical reaction. Cover with waterproof membrane as soon as possible after finish troweling. Immediate protection is required when concrete is exposed to sun or wind.

5. *Never spread dry cement on the surface to absorb excess water.* This indicates that the mix was too wet, but if necessary, remove excess water from the surface.

6. Carbon dioxide from gasoline engines or open salamanders can prevent proper hydration at the surface of curing concrete. Adequate ventilation is the answer here.

7. Fine silt or clay in the mix can work to the surface and result in dusting due to a weak mixture at the surface.

METHODS OF CURING CONCRETE

The vast majority of problems which have been experienced with concrete, such as dusting of the surface, spalling, hairline cracks, low yield strength and others, are caused by improper curing. Regardless of the quality of the mix and the care taken in handling and placing, all can be lost by improper curing.

The ultimate strength of concrete is determined by the chemical reaction process which begins as soon as water is added to the mix. The hydration process uses only about half of the water normally in the mix; the remainder must evaporate out. Consequently, a minimum of water to achieve workability is all that should be used in the mix.

Perhaps the most important rule for assurance of a good concrete, free of problems, is this: *Keep the surface moist during the entire hydration process!* If the surface becomes dry, hydration stops; and the moment hydration stops, strength gain stops. The surface must be kept moist for at least three days—and preferably for seven.

Here are five methods of protecting concrete while curing:

1. Cover with burlap and keep sprinkled. Burlap *must* be kept constantly wet. High winds and very dry weather may require sprinkling every hour or so, day and night. Under such conditions, burlap should not be used.

2. Straw is a good insulator for cold weather curing, but again, it must be kept damp. Surface of concrete should be checked every few hours.

3. Waterproof paper provides excellent protection. Extreme care must be exercised in placing to avoid damage to the surface of the concrete.

4. Plastic film is excellent. Only problem here is that the film is fragile. It must be properly weighted down and no areas of concrete can be exposed. Use opaque (black) plastic, not clear.

5. Curing compounds may be used providing no voids are left to expose any of the surface. The compound may be pigmented to facilitate full coverage. Curing compounds have no insulating value against cold weather or sun when used outside.

Caution: For use inside, be sure compounds do not contain a wax base, especially if tile floors are to be laid. On such surfaces, mastic does not dry and may work through the tile joints.

The use of a dryer concrete mix speeds the job by faster drying. Dusting of the concrete and other surface defects are avoided if a dry mix is used and worked according to proper procedures. (More of this on page 176.) If you normally sub out concrete work, it is equally important to see that the subcontractor knows what he is doing, or to find one who does. Some general contractors are doing all concrete work with their own forces to gain full control of the work and avoid the many problems which constantly arise from incorrect concrete working.

Concrete Forming

Many problems can be avoided by better form work. Why pour concrete into a rough, irregular form when it is known that the finish will not be satisfactory and will have to be ground? Grinding out irregularities in concrete can take as long as it did to place it—and possibly longer. Where grinding is necessary, the proper equipment here can pay off in a hurry with labor saved. It is a good idea to keep abreast of the latest developments in equipment.

Where smoother concrete finishes are desired, forms may be lined with one of the fiber board or form liner materials.

PROBLEM: Specifications call for a bushhammered concrete finish. This has presented quite a problem in that we are unable to get a satisfactory finish. We are in need of information on equipment and procedures.

SOLUTION: A bushhammer texture on the surface of concrete is attained by chipping away the entire surface of the concrete with a serrated faced tool known as a bushhammer. Bushhammers have been used for centuries by stone sculptors. Several styles are available from the larger suppliers of sculptor's tools and supplies.

For names and addresses of suppliers see Chapter 12.

If large areas are to be bushhammered, a pneumatic or electric power tool is recommended. A hand hammer will also be required for surfaces near the edges to prevent chipping out sections of the concrete at the edges, and for reaching areas inaccessible to the power tool.

Although a satisfactory bushhammered surface can be achieved on any good structural concrete, a more attractive texture can be attained by adhering to certain specifications. The use of colorful natural aggregate will obviously result in a more attractive surface. The concrete mix should contain at least six sacks of cement per cubic yard and the water-cement ratio should not exceed 5.5 U.S. gallons (4.3 Canadian gallons) per sack. Air-entrained concrete should be used and slump should not exceed four inches. A minimum sand content will result in a more attractive finish.

Concrete should be deposited in uniform lifts of about 16 inches in depth, and each lift vibrated to attain a uniform surface.

Warning: Concrete should not be bushhammered until it has attained a strength of 3,750 pounds per square inch or more. This will usually require about three weeks.

Chamfered corners are recommended due to the risk of chipping off edges. If sharp corners are used, be sure to texture the edges and other fragile sections with a hand hammer.

Additional information, including architects' specifications, is available from Portland Cement Association. (See Appendix.)

PROBLEM: Can you help us solve the problem of removing clay stains from concrete surfaces?

SOLUTION: Red clay usually contains a high percentage of iron oxides which contribute the red color to the clay. Iron or rust stains can be removed from masonry surfaces with the following preparation:

> Mix one part sodium citrate with six parts water.
> Add an equal volume of glycerine and mix thoroughly.
> Add whiting to form a thick paste.

Apply the mixture to the stains and allow to dry thoroughly. This may require one or two days depending upon weather conditions.

If first application does not entirely remove the stains, repeat the application until all stains have disappeared.

PROBLEM: We have a problem with rust stains on new terrazzo floors caused by nails allowed to rust on the floors. Nothing we have tried removes these rust spots.

SOLUTION: Rust stains are treated in the same manner as for the removal of clay stains described previously. If stains are not severe they may be removed with sodium hypochlorite. (Clorox is 5.25% sodium hypochlorite.)

PROBLEM: We need a solution to the problem of green stains which have appeared on the surfaces of a recently built brick structure.

SOLUTION: There are two frequent sources of green stains on masonry surfaces. One is the formation of algae which possess chlorophyll and carry on photosynthesis, making them independent of an outside food supply. Algae may be found on any type of masonry surface, trees, or even ice or snow. They will usually be found on surfaces which do not receive direct sunlight.

The other possible source of green stain is copper oxide which may be found in the vicinity of new copper. It may originate from copper flashings, facia, gutters, or other copper used in construction. New copper used in construction reacts to form copper oxide which is a blue-green coating. This may stain surfaces below or near it.

Algae and copper oxide may both be removed by the same procedure. Apply a mild solution of sodium hypochlorite to affected surfaces with a fiber scrubbing brush. If the coating is heavy, use a scraper and remove as much as possible before using the solution. One such product, known as HTH, is available from swimming pool maintenance suppliers and chemical suppliers. The HTH may be left on the surfaces since it will not damage them and will help prevent further formations.

PROBLEM: A large crack has appeared in the terrazzo work we have at the entrance to a new building. Can you give us a satisfactory method of making repairs?

SOLUTION: The cracks may be repaired by filling with an expansive cement and coloring to match the existing work. The degree of success will depend upon how well the original matrix and aggregate colors are matched. First, obtain a good white expansive cement. Two sources are Vermont Marble Company, Proctor, Vt. 05765 (their product is "Vermont White Cement"), and X-Pando Corporation, 43-15 36th St., Long Island City, N.Y. 10001 (their product is "X-Pandotite"). Building supply firms usually have one of these products.

Obtain dry colors to match those in the terrazzo. Make notes of the colors and obtain sufficient basic colors so that they can be mixed to match the aggregates in the existing work. An artist or good art student can make the match if care is exercised. The white cement will probably have to be mixed, or toned down, to match the matrix shade exactly.

Clean the crack out thoroughly with a steel brush. Then mix small batches of the cement to match the colors found in the terrazzo. Also mix a batch of white, or color to match the matrix color. Apply the different colors in balls about the same size as the original marble aggregates. Even the shape of the aggregates can be duplicated with a little effort. Force the cement well down into the crack with a tool of suitable size to enter the crack. Leave the cement patch work higher than the terrazzo surface. Do not trowel it down flush. Protect the patched area until the cement is set. After 24 hours, the patchwork can be rubbed down with a fine abrasive stone and the entire surface polished out.

Cracks and joints have been filled by this method, resulting in repairs which are almost invisible.

PROBLEMS: In several terrazzo floors we have installed, cracks have appeared. Is there any way to prevent cracking in terrazzo floors?

SOLUTION: The recommended method of terrazzo installation is the cushion type in which a one-quarter-inch sand bed is laid beneath the terrazzo and the underbed of terrazzo reinforced with wire mesh as shown in Figure 11-1. Although there is no positive assurance against cracking, this method is considered the least susceptible to cracking. If terrazzo is installed by the monolithic method, it is laid on the slab while the concrete is still green. Then if the slab cracks or moves in any way, cracks will appear in the terrazzo. In bonded-to-slab installations, divider strips must be properly laid out and strips carried over joints of stress; otherwise cracks will probably occur. Any movement or settlement in the understructure will probably result in cracks in the terrazzo.

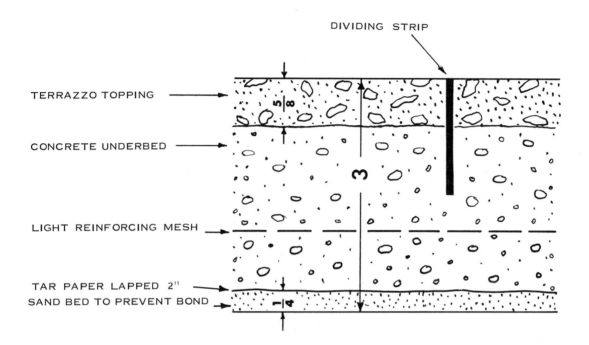

FIGURE 11-1

Cross Section of Terrazzo Floor

Detailed specifications and installation methods are available from the National Terrazzo and Mosaic Association. (See Appendix.)

PROBLEM: On a project we have underway, it is necessary to cut through three feet of high strength, reinforced concrete. To complicate the job, part of the concrete is under about three feet of water.

SOLUTION: This is an ideal job for the "Burning Bar Thermal Torch." The Burning Bar is a pipe filled with a combination of magnesium, aluminum, and steel. It is held in a special holder through which oxygen is fed at 60 to 125 pounds pressure. The equipment is simple and inexpensive. Temperatures of 4,500 to 10,000°F are produced at the end of the bar. This is sufficient heat to penetrate almost all materials. One ⅜" diameter Burning Bar will burn a 1½" to 2" hole through reinforced concrete up to three feet in thickness in six minutes.

The process is noiseless and creates very little smoke. The usual procedure for cutting concrete is to burn a series of holes and break out the remainder; however, a continuous trench can be cut through ten feet of reinforced concrete, steel or other material. Any material may be cut under water as well as in air.

The method is useful in breaking boulders. After a few holes are burned 4" to 6" deep, the intense heat cracks the rock into small segments which can then be easily handled. A heavy blow with a sledgehammer will facilitate the breaking operation.

The following guide may be used for estimating time and costs:

>Penetrating 12" reinforced concrete with 1¼" to 2" diameter hole will consume 4' to 7' of bar, 20 to 30 cu. ft. oxygen, and require about 2½ minutes of burning time.

>Penetration of metal to 12" depth consumes only 6" to 12" of bar and 15 to 20 cu. ft. oxygen. Burning time, about 1 minute.

For further information contact Burning Bar Sales Company, 6010 Yolanda Ave., Tarzana, Calif. 91356.

PROBLEM: We need some solutions to various ground water problems. We are having a problem with water filling caissons and excavations. What procedures are suggested for removal and control?

SOLUTION: Water and silt can usually be removed from the bottom of caissons by means of an air pump. Air pumps can remove as much as 50 to 60 percent solids. Where higher percentages of solids are encountered, water may be used to reduce the solid percentage by agitating the solids into solution with the water. Additional water may be introduced if necessary.

To cut off the flow of water into caissons, a steel casing with carbide tipped teeth may be sunk and cut into the bedrock by revolving the casing into the bedrock. One problem which is frequently encountered is damage to the teeth of the casing. Extreme care must be used to avoid dropping the casing against bedrock. The casing should be revolving as it approaches bottom and the main weight should be supported until it has had an opportunity to drill into the rock. Hydraulic cement can then be used after the caisson has been emptied to seal off water flow which remains.

As local conditions greatly affect methods, it is best to employ a specialist in foundation work of the type concerned. Soil solidification specialists have techniques of stabilizing the soil surrounding the caissons to effectively solidify the soil and cut off water flow.

Wellpoint Dewatering Method

Wellpoint systems may be used for ground water control, water supply, or soil stabilization. The wellpoint system can be used to keep excavations dry and to enable greater slopes. A group of small-diameter pipes are sunk into the ground surrounding the area to be excavated. These are usually spaced three to six feet apart. Water accumulating in the wellpoint pipes is then brought to the surface where it is collected in a horizontal header pipe and exhausted by a centrifugal pump with a vacuum pump assist.

The wells may be almost any depth, frequently 50 to 100 feet. Usually wellpoint systems on construction sites are temporary, but a similar system may be used for providing a permanent water supply. By sinking wells deep, surface contamination is usually bypassed.

Permanent dewatering is sometimes necessary where buildings are below grade and the wellpoint system has proved the solution to many of these problems. Recovered water may be utilized for any of numerous purposes, such as for the air conditioning system. Wellpoint systems are most effective in coarse-grain soils, such as sand or gravel, and are not so effective in fine-grain soils, such as clay or silt.

Electro-Osmosis Method

The electro-osmotic method is effective for removal of ground water in fine-grain soils, such as clay or silt. It is used to dry out areas surrounding an excavation and to stabilize the soil to permit steeper excavation slopes. This is accomplished by introducing a direct current by means of electrodes inserted into the ground. Ground water surrounding the fine grains of soil consists of three layers: an uncharged inner core, a positively charged layer, and a negatively charged layer next to the grain of soil. When a direct current potential is applied between the electrodes, the positive layer flows toward the negative electrode. The negative electrode is a pipe which is used to accumulate the water for removal by pump.

The electro-osmosis principle has been used to stabilize the soil by polarizing soil particles to increase strength.

PROBLEM: Some of the asphalt tile in a new building has become soft and we would like to know how this could happen. In some areas mastic is coming through between the joints in the tile.

SOLUTION: Some installation contractors, or individual workmen, use what they call "cut-back" mastic in which gasoline or similar solvents are used to reduce the consistency when the mastic is too thick to handle well. These solvents should not be used because they soften the tile so that it is easily damaged in normal use.

In one instance, it was found that a paint thinner had been spilled on the floor by workmen. Although it had been wiped up immediately, the thinner had been partly absorbed and had run under tiles by entering the joints. This requires only a few seconds, due to the low viscosity of the solvent.

When tile is laid on concrete the mastic solvent is normally absorbed into the concrete. Problems have been traced to paraffin-base curing agents used to retard the drying of concrete. *Paraffin-base curing agents should not be used on concrete which is to receive a mastic-laid floor covering.*

PROBLEM: Where can we find information on the manufacture of adobe brick?

SOLUTION: Adobe brick is made from earth having a mixture of clay and sand. To produce a brick of lasting quality, the earth should have a clay content of not more than 50 percent and preferably about 30 percent with the balance sand. Material for adobe brick is usually found under about 12 inches of top soil. Originally, adobe brick was made by scraping off top soil, filling the pit with water and tramping with the bare feet. Straw or some other fiber was used to reinforce the brick. Molding of the brick was done by crudely forming by hand into a brick about 4″ x 10″ x 14″ in size.

In modern construction, however, molds of wood, aluminum or other material are used to mold the brick into smaller sizes. Clay mortar was originally used to lay the brick, but modern methods utilize cement mortar. By adding an emulsion oil, or bituminous stabilizer, a "bituminous" brick or block is produced. Also, a brick called "Terrecrete" is made by adding about 5 percent cement to the adobe mix. Adobe brick up to 4″ x 8″ x 16″ is made by the extrusion process. A few manufacturers of adobe brick will be found in Southern California, Arizona, New Mexico and West Texas.

PROBLEM: Discoloration of exterior glass surfaces. Large plate glass in the front of new building has become discolored after installation. The problem is how to proceed with cleaning without risking damage to the glass.

SOLUTION: The problem can be either deposits on the surface of the glass or etching of the surface by corrosive acids or alkalies. Most deposits can be removed by the usual cleaning methods. Use a commercial or household glass cleaner or dilute solution of ammonia. Most glass cleaners contain ammonia as the major ingredient. If the stains do not yield to the cleaners, prepare the following cleaner: Mix together two tablespoons household ammonia, four tablespoons whiting, and four tablespoons denatured alcohol; then add sufficient water to make one pint. Apply this to the glass and allow to dry. Rub off and polish with a soft cloth or paper. The household product Bon Ami is excellent for this purpose. *Caution!* Many of the more abrasive cleaners will leave fine scratches.

If paint spray equipment was used near the glass, a fine over-spray may have settled on the surface. This may be removed by scraping with a new razor blade. To avoid scratches, it is best to use one of the single-edge blades by hand, being careful not to exert heavy pressure or dig in with the corner of the blade. Do not use a blunt instrument such as a putty knife. Cement should never be allowed to dry on glass, as it will almost always scratch the surface when removed. Here again, a sharp razor blade is best. Wash off all loose material before using the blade and continue washing as the cleaning progresses.

If glass does not come clean after the above efforts, scrape heavily in a less conspicuous area with the blade. If it does not come off, the glass probably has been etched. In this case, the only remedy is to polish out the surface etching. This can be done by using a power buffer with fine polishing compounds—but should be done by an expert. The buffing will remove the etched surface, but will probably leave the surface irregular and cause distorting of vision through the glass.

PROBLEM: We have had several jobs using acoustical ceiling tile and every one of them has a number of irregularities and unevenness which show up only after ceiling lights are turned on.

SOLUTION: Invariably there is going to be a certain amount of unevenness in the installation of acoustical ceiling tiles. Even small irregularities are accentuated by lights installed near the ceiling. You will find that these irregularities will not be evident if recessed lighting is used. Obviously, light rays striking the ceiling at a nearly parallel angle will greatly accentuate even the smallest irregularities. A similar condition exists where windows reach to the ceiling.

For best appearance of ceiling tiles, beveled-edge tile should be used. Also, the use of recessed fixtures will greatly enhance the appearance of the ceiling because they do not cast rays parallel to the surface.

Additional information on this subject is available from the Acoustical Society of America (see Appendix).

PROBLEM: Wood siding on an apartment building had been painted every year because of peeling paint. The best grade of paint, applied by reputable painting contractors, did not last more than one year before peeling and blistering.

SOLUTION: It was found that water had been leaking down the inside of the siding, keeping it wet almost continually. The water was absorbed through the siding from inside, keeping the siding damp except for periods during hot summer months. Each year the painting was done in the fall at a time when the siding was dry.

Several corrections were necessary:

1. Ventilation was provided at roof soffits where none existed.
2. Gutters were replaced by proper installation.
3. Roof leaks were repaired near the eaves.
4. After thorough drying of siding, old paint was removed down to bare wood, which was then primed and repainted with a top-grade paint. Paint has now been on for three years, with no evidence of peeling.

PROBLEM: What is the best solution to the problem of urine stains on carpets.

SOLUTION: Actually, urine destroys the color in carpets rather than staining it. Like all acids, it must be neutralized promptly to avoid affecting the color. On fresh urine stains, absorb as much as possible as quickly as possible with paper or towels. Then neutralize the area with a household ammonia * reduced about ten to one with plain water. Work from the edges toward the center with the ammonia solution, then clean with a mild detergent and finish with plain water. If possible, lift the carpet to allow air circulation and prevent further action by urine if still wet. If access can be had to the underside of the carpet, it should be treated in the same manner as the top side. Also neutralize the urine in the carpet pad if it is accessible.

---

* Household glass cleaner may be used.

In some cases the application of chloroform will restore the original color. Test in a very small area before proceeding.

Where these measures prove ineffective, a section of the carpet may be replaced, providing, of course, that matching material is available. Remember that material of a different dye lot is not likely to be a perfect match. This is why it is always advisable to obtain an extra yard or two of each color and pattern of carpet installed. The extra piece may be bound and used as protection to areas of high wear. This gives the piece approximately the same wear and exposure as the laid carpet, and it is always available as a repair section. If matching material is not available, the entire carpet may be dyed. A carpet cleaner and dyer should be consulted because shrinkage can be a problem.

PROBLEM: Birds roosting on buildings are causing a great nuisance. How can we get rid of them?

SOLUTION: There are several effective methods of bird control which have been developed in recent years. Method of control depends much upon conditions in each case. The National Bird Control Laboratories, 5315 West Touhy Ave., Skokie, Ill. 60076, has developed a product they call "Roost No More," which is a sticky substance applied by caulking gun or spray to surfaces where birds roost. There is the disadvantage, however, that it should not be applied where people or pets might get into it. Although it is harmless, it is unpleasant to come in contact with. On surfaces such as handrails of balconies, it is not recommended. For this type of application, nylon netting has been used to cover the roosting areas. Each problem should be studied by a specialist. Since the problem has been so universal, many more specialists are available throughout the country than in previous years.

One manufacturer of bird control devices is Bird-X, 325 West Huron St., Chicago, Ill. 60601. They make perches for sparrows, starlings and pigeons which leave a poison on the birds' feet and kills them within 12 to 48 hours. They do not die instantly, but disappear and die elsewhere. It is not harmful to songbirds since they do not normally perch on buildings. Bird-X also makes several other devices for repelling birds, including an ultrasonic generator which the human ear cannot hear, but which drives birds away. Another method used by Bird-X is a "Repulsor Ray" utilizing a rotating light beam. Birds find this frightening and blinding.

Specialists are listed in the telephone directory under "Bird Control," and others may be located by calling the best control companies.

PROBLEM: Is there any way to remove mortar stains from aluminum surfaces?

SOLUTION: Blemishes from mortar in contact with aluminum are due to corrosion of the surface from strong alkali in the mortar. Some cases we have seen were so unsightly that owners insisted on replacement of the aluminum trim. *Mortar must not be allowed to come into direct contact with aluminum!* The aluminum may be protected by coating with an alkali-resistant coating. Several such protective coatings are available from glazing contractors. Where plastic cements are used adjacent to aluminum trim, extra care should be exercised since these mortars are highly corrosive.

Aluminum should not be allowed to come into contact with wood or other absorptive materials which may become repeatedly wet unless the aluminum is well protected. Where aluminum is in contact with other types of metals, an electrolytic action takes place which results in serious corrosion. Dissimilar metals should be separated by a good quality of caulking or nonabsorptive tape or gasket material. Do not use a lead base paint on aluminum. Zinc chromate is recommended as a primer.

PROBLEM: Removal of unsightly tobacco stains from masonry surfaces.

SOLUTION: Tobacco stains can usually be removed by the following procedure: Dissolve two pounds of trisodium phosphate crystals in one gallon of hot water. Place 12 ounces of chlorinated lime in an enameled pan and dissolve by adding water slowly. Add this to the trisodium phosphate in a stoneware container and add water to make two gallons. Stir well and cover container, allowing lime to settle.

Add some of this solution to powdered talc, making a thick paste. The paste is then troweled over the stained surface. When dry, scrape off with a wooden paddle. The mixture is a strong bleaching agent and is corrosive to metals. It is an effective treatment for many other stains as well as tobacco.

Trisodium phosphate may be purchased from drug stores, and the other materials are generally available from building supply firms.

PROBLEM: How can we attain an exposed aggregate surface effect on poured-in-place concrete walls?

SOLUTION: Although exposed aggregate surfaces can best be obtained by pouring horizontally, vertical surfaces of exposed aggregate can be accomplished by using the "early wash" system. For uniformity of texture a gap-graded concrete is preferred. This method requires that the forms be stripped in about four hours when the concrete has stiffened enough to support its own weight. The surface surrounding the aggregates may then be removed with brush and water.

The following concrete mix has been used satisfactorily for a wall 12 feet high and 12 inches thick:

Portland Cement, Type 1	625 lb. per cu. yd.
Concrete sand, minus No. 4 screen	770 lb. per cu. yd.
Gravel, 1¼ to 2 in.	2230 lb. per cu. yd.
Water	300 lb. per cu. yd.
Water/cement ratio	0.48 or 5.42 gal. per sack
Sand, percent of total aggregate by volume	25.6
AEA	normal recommended amount, air not determined
Slump	not determined, estimated at 2 in.
Matrix percentage	50

The concrete was sufficiently workable to allow internal vibration and achieve good consolidation.

It is possible to paint a retarder on the concrete forms, but it is quite difficult to get an even texture in this way due to the problem of getting an even coating on the forms. The forms may absorb different amounts of the retarder in various areas and it is difficult to apply a uniform coating in the first place; consequently, penetration of the concrete surface is not consistent over all areas. Due to these problems, the early wash method is recommended.

Additional information will be found in Chapter 12 under the heading "Sources of Information on Concrete Problems."

12

Sources of Information for the Construction Industries

ASSOCIATIONS

Among the many sources of information of value to the building and construction industries are the various trade associations. Some of these are quite cooperative in providing information; others have little or no staff and can provide very little assistance. A selected list of associations dealing with building and construction activities is included in the appendix.

For associations not included in this list see the *Encyclopedia of American Associations* available at most public libraries. A copy may be obtained from the publisher, Gale Research Company, 34th Floor, Book Tower, Detroit, Mich. 48226. They also publish other directories and will provide literature upon request. This directory lists about 9,000 associations including professional societies, labor unions, Chambers of Commerce and other non-profit organizations. The directory includes trade, business, and commercial organizations, agricultural, governmental, military, legal, scientific, engineering, technical, educational, social welfare, religious, hobby and avocational organizations. It is published in a 8½" x 11" volume containing more than 700 pages.

REFERENCE SOURCES

A few of the reference sources considered of particular value to contractors are listed here. These sources are the ones most frequently used by Problem Solution Associates in answering thousands of questions and problems from their members in the building and construction industries. Many of these reference books are available in local libraries.

SOLUTIONS TO PROBLEMS IN BUILDING AND CONSTRUCTION

This is the title of an illustrated loose-leaf manual, 8½″ x 11″ page size, by Truman W. Cottom, President of Problem Solution Associates. Hundreds of solutions to the most frequently occurring and most perplexing problems that arise in building and construction are presented in simple, practical form. This manual is the result of thirty years of hard-earned experience by the author and includes the most practical and successful answers to the endless procession of problems encountered in building and construction operations.

Chapter 11 of this book is a good cross-section of the material included in this full-length treatise, so long needed by those connected with the building and construction industries.

The material is indexed and cross-indexed so that the solutions to almost any problem can be quickly located. You may be surprised to read the solutions to problems you have always thought had no solution!

Price $27.50, PSA, Post Office Box 1116, Bellaire, Tex. 77401.

BUILDING CODE ORGANIZATIONS

1. Building Officials and Code Administrators International: Basic Building Code (used in the Midwest and Northeastern states).
2. International Conference of Building Officials: Uniform Building Code (used in Western states).
3. Southern Building Code Congress: Southern Standard Building Code (in use in Southern states).
4. American Insurance Association: National Building Code (used in Eastern states).

The National Forest Products Association (see Appendix for addresses of associations) maintains a staff of building code specialists whose efforts are directed toward the development of extended uses for wood as a building material. Wood structures are restricted as to height and area, but codes have recently been relaxed, permitting larger wood structures. A check of the current situation in your area may reveal allowance of wood in larger buildings, with consequent reduction in building costs.

UNIVERSITY RESEARCH BUREAUS

Some universities are spending as much as 25 percent of their total income on research activities, and a tremendous amount of information is available to the building and construction industries from these sources. Research is continually under way in every branch of engineering and physical sciences. Most of these engineering experiment stations publish their findings in bulletins.

A description of the activities of these research institutes and bureaus will be found in the *Directory of University Research Bureaus and Institutes,* published by Gale Research Company, 34th Floor, Book Tower, Detroit, Mich. 48226. A research team somewhere is probably engaged in research on almost any problem that can arise. This directory will lead you to the source of this information. Most libraries will have a copy available for use at the library. Normally they are not for circulation.

PROBLEM SOLUTION ASSOCIATES

PSA provides a variety of information services for architects, engineers, contractors and subcontractors. Several loose-leaf manuals are available, including:

Contractor's Desk Book (new revised loose-leaf edition)
Solutions to Problems in Building and Construction
Architectural Research (manuals and services)

For additional information contact PSA, Box 1116, Bellaire, Tex. 77401.

OTHER SOURCES

Thomas Register of American Manufacturers. This is the most complete listing of manufacturers available. Published in several volumes measuring 9″ x 14″, it lists all known American manufacturers, both alphabetically and by products manufactured. A separate volume contains the index and trade name directory, a valuable part of the system. Each firm's branch offices, as well as subsidiaries and affiliates, are listed. Published annually by Thomas Publishing Company, 461 Eighth Ave., New York, N.Y. 10001, Telephone (212) 695-0500.

Sweet's Catalog Service. This is a compilation of manufacturers' catalogs, classified by type of product, indexed, and kept up to date. It is made up in three major types: Architectural, Industrial Construction, and Light Construction. F. W. Dodge Corporation, 330 W. 42nd St., New York, N.Y. 10036.

Frank R. Walker Company. This firm publishes several books for contractors and prints numerous forms for all phases of the construction business, including estimating and accounting. Many of their forms are reproduced throughout this book and in *Contractor's Desk Book*. Their estimating book contains 1,600 pages of valuable information on estimating and for general use in construction. Their book, *Practical Accounting and Cost Keeping for Contractors,* contains much valuable information as well as forms and their application. Subcontractors' and architects' systems are included. Frank R. Walker Company, 5030 N. Harlem Ave., Chicago, Ill. 60656.

Norman Foster. Norman Foster has had more than 35 years' experience in estimating and in building and construction work. He has written two books which we can recommend to everyone in building and construction. His pocket book, *Practical Tables for Building Construction,* provides a convenient source of reference material. It has many tables of sizes, weights, mixes, mathematical shortcuts, temperatures and other weather data for various locations. His other book is *Construction Estimates from Take-Off to Bid*. It has valuable tips on how to simplify the take-off and save time. It has a complete take-off and estimate for a $1,000,000 building. It includes a 36-page booklet of plan illustrations. McGraw-Hill Book Company, 1221 Avenue of the Americas, New York, N.Y. 10020.

How and Where to Find the Facts. This is an encyclopedic guide to various types of information, containing 442 pages of information for researchers. It lists names and addresses of hundreds of sources of information, and includes various research aids such as stock photo services, convention bureaus, film and television studios, and many other basic research sources. Arco Publishing Company, 480 Lexington Ave., New York, N.Y. 10017.

Guide to American Directories. A "directory of directories" describing approximately 900 directories currently being published in practically every conceivable catagory. B. Klein and Company, 27 E. 22nd St., New York, N.Y. 10010.

How and Where to Look It Up. This guide to standard sources of information is available at most libraries. It contains a wealth of information on ways to locate facts. It tells how to utilize libraries, encyclopedias, almanacs, annuals, and handbooks. It describes dictionaries, periodicals, and guides to current literature as well as information available from the various federal, state and local governments. It has sources of films, slides, and photographs. McGraw-Hill Book Company, 1221 Avenue of the Americas, New York, N.Y. 10020.

Basic Reference Sources. Published by American Library Association, 50 East Huron St., Chicago, Ill. 60611. Contains a very extensive list of sources of information.

Sources of Business Information. Lists government sources, business libraries, statistical and financial sources, and many specialized sources of information on building and materials. By Coman, published by Prentice-Hall, Inc., Englewood Cliffs, N.J. 07632.

Kelly's Directory of Merchants, Manufacturers and Shippers. The world's oldest directory of its type, it lists manufacturers and organizations of various types throughout Europe. Kelly's Directories, Ltd., 186 Strand, London W.C. 2, England. It is available in most of the larger libraries.

Standard Rate and Data Guide. The primary purpose of this guide is to furnish complete information about the various periodicals for use by advertising agencies in placing advertisements. It is useful for obtaining names and addresses of magazines. All advertising agencies subscribe to this data guide and will usually allow you to use it in their office. New issues are published frequently and your advertising agency may give you one of the older issues for use in your own office. They also publish *Consumer Markets,* which contains considerable information about geographical market areas including population, income, and sales potential. Simmons-Boardman Publishing Corporation, 30 Church St., New York, N.Y. 10007.

FEDERAL GOVERNMENT

The Federal government has such a tremendous amount of information that it would not be feasible to attempt to digest it here. The Superintendent of Documents, U.S. Printing Office, Washington, D.C., can furnish lists of available printed matter on the subject of your choice. To locate the right government official to contact on any particular problem, contact Business Service Center, U.S. Department of Commerce, Washington, D.C.

PRENTICE-HALL, INC.

Several excellent loose-leaf services are published by Prentice-Hall, Inc., Englewood Cliffs, N.J., one of the nation's largest publishers.

Personnel Management and Labor Relations

Among the many subjects dealing with personnel policies and practices are these: hiring, training, promotion, transfer, termination, working hours, vacations, holidays, absenteeism, leaves of absence, discipline, labor relations, grievances, working conditions, safety, health, job evaluation, wage and salary management, communications with employees, morale building, employee benefits, and supervision. The service includes hundreds of forms for almost every purpose related to personnel and labor. The service is thorough and complete.

Tax Digest Service

Prentice-Hall, Inc., is well known for their loose-leaf tax digest services. These are available of a size and scope to fit the type and size of the organization. Federal, state and local, as well as other taxes are covered. It is the most authoritative and complete source of information on Federal taxes available. It offers full reprints of Federal tax laws, rulings, and decisions, plus extensive expert editorial comment. It covers capital adjustments, reorganizations, stock rights, estate planning, and money-saving tax ideas. Prentice-Hall, Inc., Englewood Cliffs, N.J. 07632. The firm is represented nationally and is listed in telephone directories of major cities.

Business Services

Other services available from Prentice-Hall, Inc., are:

Prentice-Hall Executive Report—A weekly report designed to keep management personnel abreast of the latest developments and new ideas affecting the many varied aspects of business.

Consumer and Commercial Credit: Installment Sales—Covers state laws governing secured transactions. These laws include Uniform Commercial Code, Conditional Sales Contract Law, and Chattel Mortgage requirements. Also Time Sales Acts and Installment Loan coverage for all states.

Payroll Guide—Coverage of Federal, state and local laws affecting payrolls—including withholding; unemployment, old age and disability benefits; rules as to wages and hours.

Pension and Profit Sharing—Creation, installation, and administration of qualified employee pension and profit sharing plans and trusts. Included are laws, regulations, rulings, specimen plans, clauses, and forms.

Pension and Profit Sharing Forms—Model plans and numerous clauses for companies or self-employed permit tailoring to suit individual needs. Plus constant flow of new ideas and forms.

Securities Regulation—Provides complete coverage on all federal securities laws administered by the Securities and Exchange Commission, including full texts of laws, explanations and forms.

The above listed services are only a few offered by Prentice-Hall, Inc. For further information write Prentice-Hall, Inc., Englewood Cliffs, N.J. 07632.

STORMPROOFING

"Making Wood Frame Buildings More Resistant to Windstorm Damage." Request Special Interest Bulletin No. 112, American Insurance Association, 85 John St., New York, N.Y. 10038.

"Windstorm Damage Prevention." Contains suggested ways to reduce or prevent damage from windstorms, also good bibliography. American Insurance Association, 85 John St., New York, N.Y. 10038.

"Anchorage of Exterior Frame Walls." Publication No. 446, $1.50 per copy, National Academy of Sciences, National Research Council, 2101 Constitution Ave., Washington, D.C.

"Hurricane Kit" and "Tornado Kit." Two kits of information dealing with characteristics of storms. Published by Superintendent of Documents, U.S. Government Printing Office, Washington, D.C. Hurricane Kit is $0.75 and Tornado Kit is $1.00.

"Factory Roofs Need Anchorage." Write Associated Factory Mutual Life Insurance Companies, 184 High St., Boston, Mass. 02062.

COMPUTER APPLICATION TO CONSTRUCTION

In order for the contractor or designer to utilize the computer, a vast amount of information pertaining to construction costs must first be fed into the computer. To start without any data stored in memory, and no program, would make the job of getting the computer into effective use a difficult one. On the other hand, if you make use of cost data presently stored, time and costs are only a fraction of those involved in starting from scratch. One of the most complete and most current banks of construction cost data, we believe, is Management Computer Controls, Inc., 3385 Airways Blvd., Memphis, Tenn. 38130.

If you wish to locate a computer source in your own vicinity, you need only investigate the computer services organizations to find one with a satisfactory storage of the basic cost data needed. It would be advisable to check references of other contractors using the services. You will need to determine the source of the cost data and how it was developed, as well as the accuracy and currentness of the information.

VISUAL AIDS SOURCES

Visual aids include all types of photograph projection equipment and films such as motion pictures, film strips, slides. Also included are wall charts of various types, display boards, three-dimensional models, and many other devices used in educational presentations.

Training films covering many diverse subjects are available from several sources. The Jam Handy Organization has a catalog of standard training films, but their business is primarily special film-making on contract. This, of course, is quite expensive, with costs usually amounting to several thousand dollars and up into the tens of thousands for one- and two-reel subjects. Contact Jam Handy Organization, 2843 East Grand Blvd., Detroit, Mich. 48211; or 1680 Vine St., Hollywood, Calif. 90028.

Additional information on this subject will be found under "Methods Engineering Services and Equipment," which follows.

METHODS ENGINEERING SERVICES AND EQUIPMENT

Special time-lapse equipment is available from Timelapse, Inc., 1020 Corporation Way, Palo Alto, Calif. 94303.

Time-lapse equipment and services are also available from the author, Truman W. Cottom, P.O. Box 1116, Bellaire, Tex. 77401.

An excellent book on methods and equipment is *Methods Improvement for Construction Managers,* by Parker and Oglesby, published by McGraw-Hill Book Co.

CRITICAL PATH AND OTHER SCHEDULING METHODS

Several good books are available on scheduling; among them are:

Critical Path Methods in Construction Practice, 2nd ed., Antill and Woodhead, Wiley-Interscience, div. John Wiley & Sons.
Contractor's Management Handbook, O'Brien and Zilly, McGraw-Hill Book Co.

SOURCES OF SUPPLY FOR CHARTS, GRAPHS AND FORMS

Pressure sensitive transfers for lettering and illustrating charts and other visual displays will be found in your local art or drafting supply firms. They have catalogs of the available items. Naturally, no single firm will have all of the various lines available. If one source does not have what you need, just try another. Ask them to send you a catalog of stocked items. Some of the brand names are Para-tone, Chartpak, Instantype, and Para-Tipe.

Charts and components are available from Pryor Marking Products, 21 East Hubbard St., Chicago, Ill. 60611. Visual control systems, including magnetic boards, are available from Methods Research Corporation, 105 Willow Ave., Staten Island, N.Y. 10305.

Standard and special business forms for contractors are available from Frank R. Walker Company, 5030 N. Harlem Ave., Chicago, Ill. 60656.

CONCRETE FORMS AND FORM LINERS

Almost any surface effect can be obtained on concrete by using the desired forms or form liners. Smooth surfaces can be achieved by using paper or fiber forms, or by lining the forms with one of the smooth surface materials such as fiberboard, paper, plastic, metal, or similar smooth surface material.

Then, too, beautiful concrete surface treatments are obtained by using a form liner which has been molded or formed to the desired shape. Numerous sculptured effects are available in form liners.

The Thomas Register of American Manufacturers (see page 191) lists about 125 sources of forms and form liners under the heading, "Forms, Cement, Concrete, etc. Also Molds." Many textured or sculptured form liner designs can be had from Labrado Forms, 17865 Sky Park Circle, Suite 18F, Irvine, Calif. 92664.

Reusable rubber inflatable void forms are available from several sources, one of which is Elgood Hydraulics Corporation, Varick Ave. and Scholes St., Brooklyn, N.Y. 11237. These can be purchased or rented.

Seamless coated forms for round columns, piles, and similar applications are available from Sonoco Products Company, Second St., Hartsville, S.C. 29550.

SOURCES OF INFORMATION ON CONCRETE PROBLEMS

Portland Cement Association and American Concrete Institute have a vast amount of information on concrete. I have found both associations to be very cooperative on the many occasions I have had to solicit their help. Addresses of both associations will be found in the Appendix.

WELLPOINTING AND ELECTRO-OSMOSIS

Both of these subjects are treated in the McGraw-Hill *Encyclopedia of Science and Technology* which is available in most of the larger public libraries.

A discussion of wellpointing will be found in *Practices and Methods of Construction,* Steinberg and Stempel, 1957, Prentice-Hall, Inc., Englewood Cliffs, N.J. 07632.

SOURCES OF INFORMATION ON FASTENERS

The strength of construction members can be no better than the fasteners with which they are assembled. Since most failures in wood construction occur at the joints, it is advisable to consider ways of improving the assembly of wood members.

For many years, Professor E. George Stern of Virginia Polytechnic Institute, Blacksburg, Virginia, has been conducting research on fasteners and the strength of various types of wood construction. Numerous special wood research reports are available from Professor Stern's laboratory without charge.

UNUSUAL TOOLS AND EQUIPMENT

Sculpture Associates, 114 East 25th St., New York, N.Y. 10010. Bushhammers, stone carving chisels and hammers, wood carving tools, pneumatic stone carving tools, and related tools and accessories.

Burning Bar Sales Company, 6010 Yolanda Ave., Tarzana, Calif. 91356. Cutting equipment and supplies for cutting reinforced concrete, steel and other difficult materials in thicknesses up to ten feet. Methods described on page 181.

SOURCES OF INFORMATION ON MANAGEMENT DEVELOPMENT

The Management Information Center of the Administrative Management Society * reports that they service over 3,000 requests for information on professional subjects for their members annually without charge. The Society makes available to members books, magazines and reports on practically every phase of administrative management. If you are in doubt about the proper salary for a certain job, for example, the Society has published data on this subject. Courses, films, slides and manuals on management education are available. *The Administrative Management* magazine is published monthly for those who are interested in personnel and management.

INFORMATION ON OCCUPATIONAL HEALTH AND SAFETY

The Occupational Health and Safety Act of 1970 and the various revisions are published in the Federal Register, which is published weekly by the U.S. Government Printing Office, Washington, D.C. Since there have been so many revisions, however, it would be very difficult to locate all of the references. Then, too, the problem is complicated by inaccuracies, ambiguities, omissions, contradictions, and general lack of clearness. It is probably best to subscribe to one of the services available, such as Commerce Clearing House, Bureau of National Affairs, or Prentice-Hall, Inc. The value of these services is that they have attempted to digest, explain and simplify the thousands of regulations which make up the law.

Literature on Safety

The Engineering and Safety Services Department of the American Insurance Association has numerous valuable publications on almost all aspects of safety. The following subjects are published in the form of pocket-size booklets for workmen:

"Your Guide to Safety in Woodworking Shops"

"Your Guide to Safety on Construction Projects"

"Your Guide to Safe Use of Vehicular Road Construction Equipment"

---

* See Appendix for address.

"Safe Use and Care of Hand Tools"

"Safety in Painting Operations"

"Your Guide to Safety in Road Construction Work"

"Your Guide to the Safe Use of Heavy Motorized Road Building Equipment"

"Your Guide to Safety in Steel Erection Work"

I have reviewed them and they are good. Not as complete on some subjects as I would like to see, but satisfactory.

The recommended method of distribution to the workmen is in conjunction with a safety meeting at which time the salient points contained in the booklets are discussed. Try to get participation from the group. Ask them questions and get them to thinking on the subject. A supply of the desired booklets is available from American Insurance Association (see Appendix for address).

Appendix

ASSOCIATIONS

Acoustical Contractors Assn., National, 1201 Waukegan Rd., Glenview, Ill. 60025
Acoustical & Insulating Materials Assn., 111 West Washington St., Chicago, Ill. 60602
Acoustical Door Institute, c/o L.A. Ropella, U.S. Plywood, 1001 Perry St., Algoma, Wisc. 54201
Acoustical Society of America, 35 E. 45th St., New York, N.Y. 10017
Adhesives Mfg. Assn. of America, 441 Lexington Ave., New York, N.Y. 10017
Adhesive & Sealant Council (ASC), 1410 Higgins Rd., Park Ridge, Ill. 60068
Administrative Management Society, World Headquarters, Willow Grove, Pa. 19090
Aerospace Industries Assn. of America, 1725 DeSales St., N.W., Washington, D.C. 20036
Agricultural Engineers, American Society of, 1950 Niles Ave., St. Joseph, Mich. 49085
Air-Conditioning and Refrigeration Institute, 1815 N. Fort Myer Dr., Arlington, Va. 22209
Aluminum Assn., 420 Lexington Ave., New York, N.Y. 10017
Aluminum Manufacturers Assn., Architectural, 1 E. Wacker Dr., Chicago, Ill. 60601
Aluminum Siding Assn., 2217 Tribune Tower, Chicago, Ill. 60611
American Arbitration Assn., 140 W. 51st St., New York, N.Y. 10020
American Assn. of Cost Engineers, 308 Monongahela Bldg., Morgantown, W. Va. 26505
American Association of Industrial Management, 7425 Old York Rd., Melrose Park, Pa. 19126
American Assn. of Oil Well Drilling Contractors, 211 N. Ervay, Room 505, Dallas, Tex. 75201
American Assn. of Retired Persons, 1225 Connecticut Ave., N.W., Washington, D.C. 20036
American Boiler Mfrs. Assn., 1180 Raymond Blvd., Newark, N.J. 07102
American Building Contractors Assn. (ABCA), 3345 Wilshire Blvd., Los Angeles, Calif. 90005
American Ceramic Society, Inc., 4055 N. High St., Columbus, Ohio 43214
American Concrete Institute, P.O. Box 4754, Radford Sta., Detroit, Mich. 48219
American Concrete Paving Assn., 1211 W. 22nd St., Suite 727, Oak Brook, Ill. 60523
American Forest Institute, 1835 K St., N.W., Washington, D.C. 20006

American Gas Assn., 605 Third St., New York, N.Y. 10016
American Hardboard Assn., 20 N. Wacker Dr., Chicago, Ill. 60606
American Hardware Mfrs. Assn., 2130 Keith Building, Cleveland, Ohio 44115
American Home Economics Assn., 1600 20th St., N.W., Washington, D.C. 20009
American Home Lighting Institute, 360 N. Michigan Ave., Chicago, Ill. 60601
American Industrial Real Estate Assn., 5670 Wilshire Blvd., Suite 980, Los Angeles, Calif. 90036
American Institute of Architects, 1735 New York Ave., N.W., Washington, D.C. 20006
American Institute of Arch. Foundation, 6 Penn Center Plaza, Philadelphia, Pa. 19103
American Institute of Bldg. Design, Union Bank Plaza, Suite 408, 15233 Ventura Blvd., Sherman Oaks, Calif. 91403
American Institute of Consulting Engineers, 345 E. 47th St., New York, N.Y. 10017
American Institute of Interior Designers, 730 Fifth Ave., New York, N.Y. 10019
American Institute of Landscape Architects, 2721 N. Central Ave., Suite 1002, Phoenix, Ariz. 85004
American Institute of Planners, Room 800, 917 15th St., N.W., Washington, D.C. 20005
American Institute of Real Estate Appraisers, 155 E. Superior St., Chicago, Ill. 60611
American Institute of Steel Construction Inc., 101 Park Ave., New York, N.Y. 10017
American Institute of Timber Construction, 1100 17th St., N.W., Washington, D.C. 20036
American Insurance Assn., 85 John St., New York, N.Y. 10038 (Formerly Nat'l Board of Fire Underwriters)
American Iron and Steel Industries, 150 E. 42nd St., New York, N.Y. 10017
American Lumber Standards Committee, P.O. Box 1554, Rockville, Md. 20850
American Management Assn., 135 W. 50th St., New York, N.Y. 10020
American National Standards Institute, 1430 Broadway, New York, N.Y. 10018
American Paper Institute (API), 260 Madison Ave., New York, N.Y. 10016 (Formerly American Paper & Pulp Assn.)
American Pipe Fittings Assn., 60 East 42nd St., New York, N.Y. 10017
American Plywood Assn., 1119 A. St., Tacoma, Wash. 98401
American Registered Architects, Society of, 333 N. Michigan Ave., Chicago, Ill. 60601
American Road Builders Assn., 525 School St., S.W., Washington, D.C. 20024
American Society of Agricultural Engineers, 1950 Niles Ave., St. Joseph, Mich. 49085
American Society of Appraisers, 1101 17th St., N.W., Suite 305, Washington, D.C. 20036
American Society of Architectural Hardware Consultants, 104 Tiburon Blvd., Mill Valley, Calif. 94941
American Society for Church Architecture, 1700 Architects Bldg., Philadelphia, Pa. 19103
American Society of Civil Engineers, 345 East 47th St., New York, N.Y. 10017
American Society of Concrete Constructors, 2510 Dempster St., Des Plaines, Ill. 60016
American Society of Construction Inspectors, Box 1234, New York, N.Y. 10008
American Society of Golf Course Architects, 921 Port Washington Blvd., Port Washington, N.Y. 11050

APPENDIX

American Society of Heating, Refrigerating & Air Conditioning Engineers, United Engineering Center, 345 E. 47th St., New York, N.Y. 10017
American Society of Landscape Architects, 2013 Eye St., N.W., Washington, D.C. 20006
American Society of Mechanical Engineers, 345 E. 47th St., New York, N.Y. 10017
American Society of Personnel Administration, 52 E. Bridge St., Berea, Ohio 44017
American Society of Planning Officials, 1313 E. 60th St., Chicago, Ill. 60637
American Society of Real Estate Counselors, 155 E. Superior St., Chicago, Ill. 60611
American Society of Safety Engineers, 850 Busse Highway, Park Ridge, Ill. 60068
American Society for Testing Materials, 1916 Race St., Philadelphia, Pa. 19103
American Specification Institute, 134 N. LaSalle St., Chicago, Ill. 60602
American Subcontractors Assn., 402 Shoreham Bldg., Washington, D.C. 20005
American Walnut Manufacturing Assn., 666 N. Lake Shore Dr., Chicago, Ill. 60611
American Water Works Assn., 2 Park Ave., New York, N.Y. 10016
American Wood Preservers Institute, 2600 Virginia Ave., N.W., Washington, D.C. 20037
Apartment Assn. of America, 1022 15th St., N.W., Washington, D.C. 20005
Apartment Owners & Managers Assn. of America, 65 Cherry Ave., Watertown, Conn. 06975
Architects, American Institute of, 1735 New York Ave., N.W., Washington, D.C. 20006
Architects, Association of University, University Architects Office, Univ. of Mich., 326 E. Hoover, Ann Arbor, Mich. 48104
Architects, Society of American Registered, 333 N. Michigan Ave., Chicago, Ill. 60601
Architectural Aluminum Manufacturers Assn., 1 E. Wacker Dr., Chicago, Ill. 60601
Architectural Photographers Assn., 222 E. 46th St., New York, N.Y. 10017
Architectural Research, Box 1116, Bellaire, Tex. 77401
Architectural Woodwork Institute, Chesterfield House, Suite A, 5055 S. Chesterfield Rd., Arlington, Va. 22206
Asbestos Cement Products Assn., 535 5th Ave., Room 2300, New York, N.Y. 10017
Asphalt & Vinyl Asbestos Tile Institute, 101 Park Ave., New York, N.Y. 10017
Asphalt Institute, The, Asphalt Institute Bldg., College Park, Md. 20740
Asphalt Roofing Industry Bureau, 757 Third Ave., New York, N.Y. 10017
Associated Equipment Distributors, 615 West 22nd St., Oak Brook, Ill. 60521
Associated General Contractors of America, 1957 E St., N.W., Washington, D.C. 20006
Association of Asphalt Paving Technologists, 155 Experimental Engineering Bldg., University of Minnesota, Minneapolis, Minn. 55455
Association of Home Appliance Mfrs., 20 N. Wacker Dr., Chicago, Ill. 60606
Association of Oilwell Servicing Contractors, 1700 Davis Bldg., Dallas, Tex. 75202
Association of Plumbing-Heating-Cooling Contractors, National, 1016 20th St., N.W., Washington, D.C. 20036
Association, Tile Contractors of America, 112 N. Alfred St., Alexandria, Va. 22314
Association of University Architects, University Architects Office, Univ. of Mich., 326 E. Hoover, Ann Arbor, Mich. 48104

Barre Granite Assn., 51 Church St., Barre, Vt. 05641
Better Heating-Cooling Council, 393 7th Ave., New York, N.Y. 10017
Better Light, Better Sight Bureau, 750 Third Ave., New York, N.Y. 10017
Bituminous Equipment Mfg. Bureau, Suite 1700, 111 E. Wisconsin Ave., Milwaukee, Wisc. 53202
Bituminous Pipe Institute, 333 N. Michigan Ave., Chicago, Ill. 60601
Blue Print & Allied Industries, International Assn., 33 E. Congress Parkway, Chicago, Ill. 60605
Boiler & Radiator Mfrs., Institute of, 393 Seventh Ave. (10th Floor), New York, N.Y. 10001
Boiler Mfrs. Assn., American, 1180 Raymond Blvd., Newark, N.J. 07102
Building Officials Conference of America, Inc., 1313 E. 60th St., Chicago, Ill. 60637
Building Owners & Managers, International Assn. of, 134 S. LaSalle St., Chicago, Ill. 60603
Building Research Advisory Board, 2101 Constitution Ave., Washington, D.C. 20418
Building Stone Institute, 420 Lexington Ave., New York, N.Y. 10017
Building Waterproofers Assn., Inc., 60 E. 42nd St., New York, N.Y. 10017
Bureau of Explosives, 2 Pennsylvania Plaza, New York, N.Y. 10001

California Redwood Assn., 617 Montgomery St., San Francisco, Calif. 94111
Carpet & Rug Institute, 208 W. Cuyler St., P.O. Box 8, Dalton, Ga. 30720
Cast Iron Pipe Research Assn., 1211 W. 22nd St., Suite 323, Oak Brook, Ill. 60521
Cedar Assn., Western Red & Northern White, P.O. Box 2786, New Brighton, Minn. 55112
Cedar Lumber Assn., Western Red, 707 Joseph Vance Bldg., Seattle, Wash, 98101
Cement Assn., Portland, Old Orchard Rd., Skokie, Ill. 60076
Ceramic Society, Inc., American, 4055 N. High St., Columbus, Ohio 43214
Church Architecture, American Society for, 1700 Waukegan Rd., Glenview, Ill. 60025
Chamber of Commerce of the U.S., 1615 H St., N.W., Washington, D.C. 20006
Civil Engineers, American Society of, 345 East 47th St., New York, N.Y. 10017
Clay Flue Lining Institute, P.O. Box 152, Perkasie, Pa. 18944
Clay Pipe Institute, National, 350 W. Terra Cotta Ave., Crystal Lake, Ill. 60014
Clay Products Institute, Structural, 1750 Old Meadow Rd., McLean, Va. 22101
Clay & Slate Institute, Expanded Shale, 1041 National Press Bldg., Washington, D.C. 20004
Committee for National Land Development Policy, 919 N. Michigan Ave., Chicago, Ill. 60611
Concrete Assn., National Ready Mixed, 900 Spring St., Silver Spring, Md. 20910
Concrete Institute, American, P.O. Box 4754, Radford Station, Detroit, Mich. 48219
Concrete Institute, Prestressed, 205 W. Wacker Drive, Chicago, Ill. 60606
Concrete Joint Institute, Two Kimball, Elgin, Ill. 60120
Concrete Masonry Assn., National, 2009 14th St., North, Arlington, Va. 22201
Concrete Pipe Assn., 1501 Wilson Bldg., Arlington, Va. 22209

APPENDIX

Concrete Reinforcing Steel Institute, 228 N. LaSalle St., Chicago, Ill. 60601
Contracting Plasterers' & Latherers' International Assn., 20 E St., N.W., Washington, D.C. 20001
Contractors, National Assn. of Elevator, 2772 S. Randolph St., Arlington, Va. 22206
Contractors Pump Bureau, 20th & E Streets, N.W., Washington, D.C. 20006
Construction Industry Collective Bargaining Commission, Constitution Ave. & 14th St., N.W., Washington, D.C. 20210
Construction Industry Mfrs. Assn., Suite 1700 Marine Plaza, 111 E. Wisconsin Ave., Milwaukee, Wisc. 53202
Construction Specifications Institute, 1150 Seventeenth St., N.W., Washington, D.C. 20036
Construction Surveyors Institute, 420 Lexington Ave., New York, N.Y. 10017
Construction Writers Assn., 601 13th St., N.W., Room 202, Washington, D.C. 20005
Constructors' Assn., National, 801 Continental Bldg., Washington, D.C. 20005
Cooling Tower Institute, 3003 Yale Ave., Houston, Tex. 77018
Copper Development Assn., 405 Lexington Ave., 57th Floor, New York, N.Y. 10017
Copper Assn., United States, 50 Broadway, New York, N.Y. 10004
Copper Institute, 50 Broadway, New York, N.Y. 10004
Copper Research Assn., International, 1271 Avenue of the Americas, New York, N.Y. 10020
Corrosion Engineers, National Assn. of, 2400 W. Loop South, Houston, Tex. 77027
Council of Educational Facility Planners, 29 West Woodruff Ave., Columbus, Ohio 43201
Council of Mechanical Specialty Contracting Industries, 1611 N. Kent St., Suite 200, Arlington, Va. 22209
Council of Profit Sharing Industries, 29 N. Wacker Dr., Chicago, Ill. 60606
CPM Consultants, Div. of Society for Advancement of Management, 26 Fellowship Rd., Cherry Hill, N.J. 08034
Craftsman's Guild (Mobile Homes), P.O. Box 915, Elkhart, Ind. 46514
Cypress Mfrs. Assn., Southern, P.O. Box 16413, 1640 West Rd., Jacksonville, Fla. 32216

Design Legislation, National Committee for Effective, Suite 2700, 200 E. 42nd St., New York, N.Y. 10017
Designers, American Institute of Interior, 730 Fifth Ave., New York, N.Y. 10019
Designer's Society of America, Industrial, 60 West 55th St., New York, N.Y. 10019 (Replaces Industrial Designer's Institute)
Distribution Contractors, 531 Harvard Tower, 4815 S. Harvard, Tulsa, Okla. 74135
Door Institute, Acoustical, c/o L. A. Ropella, U.S. Plywood, 1001 Perry St., Algoma, Wisc. 54201
Door Institute, Steel, 2130 Keith Bldg., Cleveland, Ohio 44115
Door Operator & Remote Controls Mfrs. Assn., 110 N. Wacker Drive, Chicago, Ill. 60606
Douglas Fir Plywood Assn. (Now American Plywood Assn.)

Edison Electric Institute, 750 Third Ave., New York, N.Y. 10017
Electrical & Electronics Engineers, Institute of, 345 E. 47th St., New York, N.Y. 10017
Electrical Mfrs. Assn., National, 155 E. 44th St., New York, N.Y. 10017
Electric Institute, Edison, 750 Third Ave., New York, N.Y. 10017
Elevator Contractors, National Assn. of, 2772 Randolph St., Arlington, Va. 22206
Elevator Industry Inc., National, 101 Park Ave., New York, N.Y. 10017
Engineers, American Institute of Consulting, 345 E. 47th St., New York, N.Y. 10017
Expanded Shale, Clay & Slate Institute, 1041 National Press Bldg., Washington, D.C. 20004
Expansion Joint Mfrs. Assn., 331 Madison Ave., New York, N.Y. 10017
Explosives, Institute of Makers, 420 Lexington Ave., New York, N.Y. 10017

Facing Tile Institute, 333 N. Michigan Ave., Chicago, Ill. 60601
Factory Mutual System (Fire Ins.), 1151 Boston-Providence Turnpike, Norwood, Mass. 02062
Fasteners Institute, Industrial, 1505 East Ohio Bldg., Cleveland, Ohio 44114
Felt Manufacturers Council, c/o Northern Textile Assn., 211 Congress St., Boston, Mass. 02110
Fine Hardwoods Assn., 666 Lake Shore Drive, Chicago, Ill. 60611
Fire Equipment Distributors, National Assn. of, 604 Davis St., P.O. Box 1406, Evanston, Ill. 60204
Fire Equipment Mfrs. Assn., 604 Davis St., P.O. Box 1406, Evanston, Ill. 60204
Fire Protection Assn., National, 60 Batterymarch St., Boston, Mass. 02110
Fire Underwriters, National Board of (Now American Insurance Assn.)
Fireplace Assn. of America, 435 N. Michigan Ave., Suite 2700, Chicago, Ill. 60611
Flooring Division, Vinyl & Rubber, Rubber Mfrs. Assn., 444 Madison Ave., New York, N.Y. 10022
Flooring Institute of America, Wood, 201 N. Wells St., Chicago, Ill. 60606
Flooring Mfrs. Assn., Maple, 424 Washington Ave., Oshkosh, Wisc. 54901
Flooring Mfrs. Assn., National Oak, 814 Sterick Bldg., Memphis, Tenn. 38103
Forest Industries Assn., Western, 1500 Southwest Taylor, Portland, Ore. 92205
Forest Institute, American, 1816 K St., N.W., Washington, D.C. 20006
Forest Products Research Society, 2801 Marshall Court, Madison, Wisc. 53705
Fuel Institute, National Oil, 60 E. 42nd St., New York, N.Y. 10017

Gas Appliance Mfrs. Assn., 60 E. 42nd St., New York, N.Y. 10017
Gas Assn., American, 605 Third Ave., New York, N.Y. 10016
Gas Assn., National LP, 79 W. Monroe St., Chicago, Ill. 60603
General Contractors of America, Associated, 1957 E St., N.W., Washington, D.C. 20006
Glass Assn. of America, Stained, 3600 University Dr., Fairfax, Va. 22030
Glass Mfrs. Assn., Sealed Insulating, 2217 Tribune Tower, Chicago, Ill. 60611
Glue Mfrs. Inc., National Assn. of, 663 Fifth Ave., New York, N.Y. 10022

APPENDIX 205

Granite Assn., Barre, 51 Church St., Barre, Vt. 05641
Granite Quarries Assn., National Building, Concord, N.H. 03301
Ground Water Institute, P.O. Box 981, Minneapolis, Minn. 55440
Gypsum Association, 201 N. Wells St., Chicago, Ill. 60606
Gypsum Roof Deck Foundation, 1201 Waukegan Rd., Glenview, Ill. 60025

Hardboard Assn., American, 20 N. Wacker Dr., Chicago, Ill. 60606
Hardware Assn., National Builders, 1290 Avenue of the Americas, New York, N.Y. 10019
Hardware Assn., National Retail, 964 N. Pennsylvania St., Indianapolis, Ind. 46204
Hardware Consultants, American Society of Architectural, 104 Tiburon Blvd., Mill Valley, Calif. 94941
Hardware Mfrs. Assn., Drapery, 331 Madison Ave., New York, N.Y. 10017
Hardwood & Pine Mfrs. Assn., Northern, Northern Bldg., Suite 207, Green Bay, Wisc. 54178
Hardwood Dimension Mfrs. Assn., 3813 Hillsboro Rd., Nashville, Tenn. 37215
Hardwood Lumber Assn., National, 59 E. Van Buren St., Chicago, Ill. 60605
Hardwood Lumber Mfrs. Assn., Southern, 805 Sterick Bldg., Memphis, Tenn. 38103
Hardwood Mfrs., Appalachian, (Lumber), 414 Walnut St., Cincinnati, Ohio 45202
Hardwood Plywood Mfrs. Assn., P.O. Box 6246, Arlington, Va. 22208
Hardwoods Assn., Fine, 666 Lake Shore Dr., Chicago, Ill. 60611
Heating & Air Conditioning Wholesalers Assn., North American, 1200 West Fifth St., Columbus, Ohio 43212
Heating Assn., International District, 5940 Baum Square, Pittsburgh, Pa. 15206
Heating-Cooling Council, Better, 393 7th Ave., New York, N.Y. 10001
Heating, Refrigerating & Air Conditioning Engineers, American Society of, United Engineering Center, 345 E. 47th St., New York, N.Y. 10017
Highway Officials, American Assn. of State, 341 National Press Bldg., Washington, D.C. 20004
Highway Research Board, 2101 Constitution Ave., N.W., Washington, D.C. 20418
Home Appliance Mfrs., Assn. of, 20 N. Wacker Drive, Chicago, Ill. 60606
Home Builders of the U.S., National Assn. of, 1625 L St., N.W., Washington, D.C. 20036
Home Economics Assn., American, 1600 20th St., N.W., Washington, D.C. 20009
Home Improvement Council, National, 11 E. 44th St., New York, N.Y. 10017
Home Lighting Institute, American, 360 N. Michigan Ave., Chicago, Ill. 60601
Home Mfrs. Assn., 1625 L St., N.W., Washington, D.C. 20036
Home Ventilating Institute, 1108 Standard Bldg., Cleveland, Ohio 44113
Hotel & Motel Assn., American, 221 West 57th St., New York, N.Y. 10019
Housing & Redevelopment Officials, National Assn. of, 2600 Virginia Ave., Washington, D.C. 20037
Housing, Assn. for Middle Income, 217 Park Row, New York, N.Y. 10038
Housing Conference, National, 1250 Connecticut Ave., N.W., Washington, D.C. 20036

Housing, National Committee Against Discrimination in, 1865 Broadway, New York, N.Y. 10023

Illuminating Engineering Research Institute, 345 E. 47th St., New York, N.Y. 10017
Illuminating Engineering Society, 345 East 47th St., New York, N.Y. 10017
Imported Hardwood Products Assn., World Trade Center, Ferry Bldg., San Francisco, Calif. 94111
Incinerator Institute of America, 60 E. 42nd St., Suite 1914, New York, N.Y. 10017
Indiana Limestone Institute of America, P.O. Box 489, Bloomington, Ind. 47401
Industrial Designers Society of America, 60 West 55th St., New York, N.Y. 10019
Industrial Fasteners Institute, 1505 E. Ohio Bldg., Cleveland, Ohio 44114
Industrial Forestry Assn., 1410 S.W. Morrison St., Portland, Ore. 97205
Institute of Architects Foundation, American, 6 Penn Center Plaza, Philadelphia, Pa. 19103
Institute of Boiler & Radiator Mfrs., 393 Seventh Ave., 10th Floor, New York, N.Y. 10001
Institute of Electrical & Electronics Engineers, 345 E. 47th St., New York, N.Y. 10017
Institute of Fireplace Equipment Mfrs., 333 N. Michigan Ave., Chicago, Ill. 60601
Institute of Makers of Explosives, 420 Lexington Ave., New York, N.Y. 10017
Institute of Masonry Research, 9013 Old Hartford Rd., Baltimore, Md. 21234
Institute of National Oil Fuel, 60 East 42nd St., New York, N.Y. 10017
Institute of Store Planners, 415 East 53rd St., New York, N.Y. 10022
Insulation Distributor-Contractors, National Assn., 8226 Fenton St., Silver Spring, Md. 20910
Insulation Mfrs. Assn., National, 441 Lexington Ave., New York, N.Y. 10017
International Assn. of Blueprints & Allied Industries, 33 East Congress Parkway, Chicago, Ill. 60605
International Assn. of Electrical Inspectors, 201 E. Erie St., Chicago, Ill. 60611
International Assn. of Plumbing & Mech. Officials, 5032 Alhambra Ave., Los Angeles, Calif. 90032
International Builders Exchange Executives, 1175 Dublin Rd., Columbus, Ohio 43215
International Conference of Building Officials, 50 S. Los Robles, Pasadena, Calif. 91101
International Contracting Plasterers' & Latherers' Assn., 20 E St., N.W., Washington, D.C. 20001
International Copper Research Assn., 1271 Avenue of the Americas, New York, N.Y. 10020
International Council of Shopping Centers, 445 Park Ave., New York, N.Y. 10022
International Stained Glass Assn., 15 Prince St., Paterson, N.J. 07505
Iron & Steel Engineers, Association of, 1010 Empire Bldg., Pittsburgh, Pa. 15222
Iron & Steel Institute, American, 150 E. 42nd St., New York, N.Y. 10017

Joint Industry, Board of the Electrical Industry, 158-11 Jewel Ave., Flushing, Queens, N.Y. 11365

APPENDIX

207

Joint Institute, Concrete, Two Kimball St., Elgin, Ill. 60120
Joints, Research Council on Riveted & Bolted, c/o Industrial Fasteners Institute, 1517 Terminal Tower, Cleveland, Ohio 44113
Joist Institute, Steel, 2001 Jefferson Davis Highway, Arlington, Va. 22202

Kitchen Cabinet Assn., National, 334 E. Broadway, Louisville, Ky. 40202

Land Development Policy, Committee for National, 919 N. Michigan Ave., Chicago, Ill. 60611
Land Institute, Urban, 1200 18th Street, N.W., Washington, D.C. 20036
Landscape Architects, American Society of, 2013 Eye St., N.W., Washington, D.C. 20036
Lathing & Plastering, National Bureau for, 938 K St., N.W., Washington, D.C. 20001
Lead Industries Assn. Inc., 292 Madison Ave., New York, N.Y. 10017
Lighting Institute, American Home, 360 N. Michigan Ave., Chicago, Ill. 60601
Lighting Protection Assn., United, P.O. Box 462, Ithaca, N.Y. 14850
Lightweight Aggregate Producers Assn., B&B Bldg., 546 Hamilton St., Allentown, Pa. 18101
Lime Assn., National, 4000 Brandywine St., N.W., Washington, D.C. 20016
Limestone Institute of America, Indiana, P.O. Box 489, Bloomington, Ind. 47401
Limestone Institute, National, 702 H St., N.W., Washington, D.C. 20001
LP-Gas Association, National, 70 W. Monroe St., Chicago, Ill. 60603
Lumber & Building Material Dealers Assn., National, 302 Ring Building, Washington, D.C. 20036
Lumber Assn., National-American Wholesale, 180 Madison Ave., New York, N.Y. 10016

Mahogany Assn., Philippine, P.O. Box 3362, Tacoma, Wash. 98499
Management Assn., American, 135 West 50th St., New York, N.Y. 10020
Maple Flooring Mfrs. Assn., 424 Washington Ave., Oshkosh, Wisc. 54901
Marble Institute of America, 848 Pennsylvania Bldg., Washington, D.C. 20004
Mason Contractors Assn. of America, 208 S. LaSalle St., Chicago, Ill. 60604
Masonry Assn., National Concrete, 2009 14th St., North, Arlington, Va. 22201
Masonry Research, Institute of, 9013 Old Hartford Rd., Baltimore, Md. 21234
Material Handling Equipment Distribution Assn., 20 N. Wacker Dr., Chicago, Ill. 60606
Mechanical Contractors Assn. of America, 2 Pennsylvania Plaza, Suite 1950, New York, N.Y. 10001
Mechanical Engineers, American Society of, 345 East 47th St., New York, N.Y. 10017
Metal Building Dealers Assn., 221 N. LaSalle St., Chicago, Ill. 60601
Metal Building Mfrs. Assn., 2130 Keith Bldg., Cleveland, Ohio 44115
Metal Lath Assn., 221 N. LaSalle St., Chicago, Ill. 60601

Metal Mfrs., National Assn. of Architectural, 228 N. LaSalle St., Chicago, Ill. 60601
Mineral Wood Insulation Assn., National, 211 E. 51st St., New York, N.Y. 10022
Mirror Mfrs., National Assn. of, 1225 19th St., N.W., Room 807, Washington, D.C. 20036
Mobile Homes Mfrs. Assn., 20 N. Wacker Drive, Chicago, Ill. 60606
Mobile Housing Assn. of America, 39 S. LaSalle St., Chicago, Ill. 60603
Model Code Standardization Council, 50 S. Los Robles Blvd., Pasadena, Calif. 91101
Moles, The, Biltmore Hotel, Madison Ave. & 43rd St., New York, N.Y. 10017 (below grade construction)
Monument Builders of North America, 1612 Central St., Evanston, Ill. 60201
Mortgage Bankers Assn. of America, 1707 H St., N.W., Washington, D.C. 20006
Motel Assn. of America, 1025 Vermont Ave., N.W., Washington, D.C. 20005

National Acoustical Contractors Assn., 1201 Waukegan Rd., Glenview, Ill. 60025
National-American Wholesale Lumber Assn., 180 Madison Ave., New York, N.Y. 10016
National Apartment Assn., 5050 Westheimer, Suite 313, Houston, Tex. 77027
National Asphalt Pavement Assn., 6715 Kenilworth Ave., Riverdale, Md. 20840
National Assn. Insulation Distributor-Contractors, 8226 Fenton St., Silver Spring, Md. 20910
National Assn. of Accountants, 505 Park Ave., New York, N.Y. 10022
National Assn. of Architectural Metal Mfrs., 228 N. LaSalle St., Chicago, Ill. 60601
National Assn. of Building Owners & Managers, 134 S. LaSalle St., Chicago, Ill. 60603
National Assn. of Carpet Specialists, c/o Harold A. Sakayon, 1430 K St., N.W., Washington, D.C. 20005
National Assn. of Corrosion Engineers, 2400 West Loop South, Houston, Tex. 77027
National Assn. of Elevator Contractors, 2772 S. Randolph St., Arlington, Va. 22206
National Assn. of Floor Covering Installers, 3910 Georgia Ave., N.W., Washington, D.C. 20011
National Assn. of Home Builders of the United States, 1625 L St., N.W., Washington, D.C. 20036
National Assn. of Housing Cooperatives, 1012 14th St., N.W., Washington, D.C. 20005
National Assn. of Housing & Redevelopment Officials, 2600 Virginia Ave., Washington, D.C. 20037
National Assn. of Lighting Maintenance Contractors, 5650 E. Evans Ave., #27, Denver, Colo. 80222
National Assn. of Mfrs., 227 Park Ave., New York, N.Y. 10017
National Assn. of Mirror Mfrs., 1225 19th St., N.W., Room 807, Washington, D.C. 20036
National Assn. of Plastic Fabricators, 4720 Montgomery Lane, Bethesda, Md. 20014
National Assn. of Plumbing-Heating-Cooling Contractors, 1016 20th St., N.W., Washington, D.C. 20036
National Assn. of Real Estate Boards, 155 E. Superior, Chicago, Ill. 60611
National Assn. of River & Harbor Contractors, 3900 N. Charles St., Baltimore, Md. 21218

APPENDIX

National Assn. of Women in Construction (NAWIC), 1000 Vermont Ave., N.W., Washington, D.C. 20005
National Automatic Sprinkler & Fire Control Assn., 2 Holland Ave., White Plains, N.Y. 10603
National Better Business Bureau, 230 Park Ave., New York, N.Y. 10017
National Builders Hardware Assn., 1290 Avenue of the Americas, New York, N.Y. 10019
National Building Material-Distributors Assn., 221 N. LaSalle, Chicago, Ill. 60604 (Formerly National Plywood Dist. Assn.)
National Building Products Assn., 120-44 Queens Blvd., Kew Gardens, N.Y. 11415
National Bureau for Lathing & Plastering, 938 K St., N.W., Washington, D.C. 20001
National Business Forms Assn., 300 N. Lee, Alexandria, Va. 22314
National Certified Pipe Welding Bureau, 2 Pennsylvania Plaza, Suite 1950, New York, N.Y. 10001
National Commercial Refrigerator Sales Assn., 1900 Arch St., Philadelphia, Pa. 19103
National Committee for Effective Design Legislation, 200 E. 42nd St., Suite 2700, New York, N.Y. 10017
National Concrete Masonry Assn., 2009 14th St., North, Arlington, Va. 22201
National Constructors Assn., 801 Continental Bldg., Washington, D.C. 20005
National Contract Management Assn., 1626 Centinela Bl., Inglewood, Calif. 90302
National Council of Acoustical Consultants, 2261 Winthrop Rd., Trenton, Mich. 48183
National Council of Specialty Cont. Assn., 1022 15th St., N.W., Washington, D.C. 20005
National Crushed Stone Assn., 1415 Elliott Pl., N.W., Washington, D.C. 20007
National Electrical Contractors Assn., 1730 Rhode Island Ave., N.W., Washington, D.C. 20036
National Electrical Mfrs., Assn., 155 E. 44th St., New York, N.Y. 10017
National Elevator Industry Inc., 101 Park Ave., New York, N.Y. 10017
National Environmental Systems Contractors Assn., 221 N. LaSalle St., Chicago, Ill. 60601
National Fire Protection Assn., 60 Batterymarch St., Boston, Mass. 22110
National Forest Products Assn., 1619 Mass. Ave., N.W., Washington, D.C. 20036
National Glass Dealers Assn., 1000 Connecticut Ave., Suite 610, Washington, D.C. 20036
National Hardwood Lumber Assn., 50 E. Van Buren St., Chicago, Ill. 60605
National Home Fashions League, 2100 Stemmons Freeway, Dallas, Tex. 75207
National Home Improvement Council, 11 E. 44th St., New York, N.Y. 10017
National Housing Conference, 1250 Connecticut Ave., N.W., Suite 632, Washington, D.C. 20036
National Insulation Mfrs. Assn., 441 Lexington Ave., New York, N.Y. 10017
National Kitchen Cabinet Assn., 334 E. Broadway, Louisville, Ky. 40202
National Landscape Nurserymen's Assn., 832 Southern Bldg., Washington, D.C. 20005
National Lime Assn., 4000 Brandywine St., N.W., Washington, D.C. 20016
National L-P Gas Assn., 79 W. Monroe St., Chicago, Ill. 60603
National Lumber & Building Material Dealers Assn., 302 Ring Bldg., Washington, D.C. 20036

National Management Assn., 333 West First St., Dayton, Ohio 45402
National Mineral Wool Insulation Assn., 211 E. 51st St., New York, N.Y. 10022
National Oak Flooring Mfrs. Assn., 814 Sterick Building, Memphis, Tenn. 38103
National Oil Fuel Institute, 60 East 42nd St., New York, N.Y. 10017
National Paint, Varnish & Lacquer Assn., 1500 Rhode Island Ave., N.W., Washington, D.C. 20005
National Parking Assn., 1101 17th St., N.W., Suite 906, Washington, D.C. 20036
National Particleboard Assn., 711 14th St., N.W., Washington, D.C. 20005
National Pest Control Assn., 250 W. Jersey St., Elizabeth, N.J. 07202
National Quartz Producers Council, (NQPC-Stone), P.O. Box 771, Golden, Colo. 80401
National Ready Mixed Concrete Assn., 900 Spring St., Silver Spring, Md. 20910
National Remodelers Assn., 50 E. 42nd St., New York, N.Y. 10017 (Formerly National Established Repair Service & Improvement Contractors Assn.)
National Retail Hardware Assn., 964 N. Pennsylvania St., Indianapolis, Ind. 46204
National Roofing Contractors Assn., 1515 Harlem Ave., Oak Park, Ill. 60302
National Safety Council, 425 N. Michigan Ave., Chicago, Ill. 60611
National Sand & Gravel Assn., 900 Spring St., Silver Spring, Md. 20910
National Sash & Door Jobbers Assn., 20 N. Wacker Dr., Chicago, Ill. 60606
National Slate Assn., 455 W. 23rd St., New York, N.Y. 10033
National Society of Professional Engineers, 2029 K St., N.W., Washington, D.C. 20006
National Spray Painting & Finishing Equipment Assn., P.O. Box 913, Toledo, Ohio 43601
National Swimming Pool Institute, 2000 K St., N.W., Washington, D.C. 20006
National Terrazzo & Mosaic Assn., 716 Church St., Alexandria, Va. 22314
National Truck Leasing Assn., 23 E. Jackson Blvd., Chicago, Ill. 60604
National Utility Contractors Assn., 815 15th St., N.W., Washington, D.C. 20005
National Water Well Assn., 88 E. Broad St., Columbus, Ohio 43215
National Woodwork Mfrs. Assn., 400 N. Madison St., Chicago, Ill. 60606
Northern Hardwood & Pine Mfrs. Assn., Northern Bldg., Suite 207, Green Bay, Wisc. 54178
Northwest Hardwood Assn., 747 St. Helens, Room 402, Tacoma, Wash. 98402

Oak Flooring Mfrs. Assn., National, 814 Sterick Bldg., Memphis, Tenn. 38103
Oil Fuel Institute, National, 60 East 42nd St., New York, N.Y. 10017

Pacific Lumber Inspection Bureau, 5557 White-Henry-Stuart Bldg., Seattle, Wash. 98101
Paint Research Institute, c/o Dept. of Chemistry, Kent State University, Kent, Ohio 44240
Paint & Wallpaper Assn. of America, 2101 S. Brentwood, St. Louis, Mo. 63144
Paint, Varnish & Lacquer Assn., National, 1500 Rhode Island Ave., N.W., Washington, D.C. 20005

Painting & Decorating Contractors of America, 2625 West Peterson Ave., Chicago, Ill. 60645
Painting & Finishing Equipment Assn., National Spray, P.O. Box 913, Toledo, Ohio 43601
Painting Council, Steel Structures, 4400 Fifth Ave., Pittsburgh, Pa. 15213
Particleboard Assn., National, 711 14th St., N.W., Washington, D.C. 20005
Perlite Institute Ins., 45 W. 45th St., New York, N.Y. 10036
Pest Control Assn., National, 250 W. Jersey St., Elizabeth, N.J. 07202
Petroleum Equipment Contractors Assn., 474 Schiller St., Elizabeth, N.J. 07206
Philippine Mahogany Assn., P.O. Box 3362, Tacoma, Wash. 98499
Pine Assn., Southern, P.O. Box 52648, National Bank of Commerce Bldg., New Orleans, La. 70150
Pine Woodwork, Ponderosa, 39 S. LaSalle St., Chicago, Ill. 60603
Pipe Fabrication Institute, 992 Perry Highway, Pittsburgh, Pa. 15237
Pipe Fittings Assn., American, 60 E. 42nd St., New York, N.Y. 10017
Pipe Institute, Bituminous, 333 N. Michigan Ave., Chicago, Ill. 60601
Pipe Institute, Cast Iron Soil, 2029 K St., N.W., Washington, D.C. 20006
Pipe Institute, Plastic, 250 Park Ave., New York, N.Y. 10017
Pipe Line Construction Assn., 2800 Republic National Bank Bldg., Dallas, Tex. 75201
Pipe Research Assn., Cast Iron, 1211 W. 22nd St., Suite 323, Oak Brook, Ill. 60521
Planners, American Institute of, Room 800, 15th St., N.W., Washington, D.C. 20005
Planning Officials, American Society of, 1313 E. 60th St., Chicago, Ill. 60637
Plasterers' & Latherers' Assn., International Contracting, 20 E Street, N.W., Washington, D.C. 20001
Plastic Industry, Society of the, 250 Park Ave., New York, N.Y. 10017
Plastics in Construction Council, c/o Jay G. Somers, Consoweld Corp., Wisconsin Rapids, Wisc. 54494
Plastics Pipe Institute, 250 Park Ave., New York, N.Y. 10017
Plumbing Brass Institute, 221 N. LaSalle St., Chicago, Ill. 60601
Plumbing & Drainage Institute (PDI), 1018 N. Austin Blvd., Oak Park, Ill. 60302
Plumbing-Heating-Cooling Contractors, National Assn. of, 1016 20th St., N.W., Washington, D.C. 20036
Plumbing-Heating-Cooling Information Bureau, 35 E. Wacker Dr., Chicago, Ill. 60601
Plywood International, Box 1337, Tacoma, Wash. 98401 (Formerly Pacific Forest Industries)
Plywood Research Foundation, 620 East 26th St., Tacoma, Wash. 98421
Ponderosa Pine Woodwork, 39 S. LaSalle St., Chicago, Ill. 60603
Porcelain Enamel Institute, 1900 L St., N.W., Washington, D.C. 20036
Portland Cement Assn., Old Orchard Rd., Skokie, Ill. 60076
Power & Communication Contractors Assn., Box 306, Early, Ia. 50535
Power Tool Institute, P.O. Box 1406, 604 Davis St., Evanston, Ill. 60204
Pressure Sensitive Tape Council, 1201 Waukegan Rd., Glenview, Ill. 60025
Prestressed Concrete Institute, 205 W. Wacker Dr., Chicago, Ill. 60606
Problem Solution Associates, Box 1116, Bellaire, Tex. 77401

Professional Engineers, National Society of, 2029 K St., N.W., Washington, D.C. 20006
Producers Council, 1717 Massachusetts Ave., N.W., Washington, D.C. 20036
Profit Sharing Research Foundation, 1718 Sherman Ave., Evanston, Ill. 60201
Pump Mfrs. Assn., Sump, 221 N. LaSalle St., Chicago, Ill. 60601

Real Estate Appraisers, American Institute of, 155 E. Superior St., Chicago, Ill. 60611
Real Estate Appraisers, Society of, 7 South Dearborn St., Chicago, Ill. 60603
Real Estate Boards, National Assn. of, 155 E. Superior, Chicago, Ill. 60611
Red Cedar Shingle & Handsplit Shake Bureau, 5510 White Bldg., Seattle, Wash. 98101
Research Advisory Board, Building, 2101 Constitution Ave., Washington, D.C. 20418
Reinforced Concrete Research Council, 5420 Old Orchard Rd., Skokie, Ill. 60078
Roof Deck Foundation, Gypsum, 1201 Waukegan Road, Glenview, Ill. 60025
Roofing Contractors Assn., National, 1515 N. Harlem Ave., Oak Park, Ill. 60302
Roofing Industry Bureau, Asphalt, 737 Third Ave., New York, N.Y. 10017

Safety Council, National, 425 N. Michigan Ave., Chicago, Ill. 60611
Sand & Gravel Assn., National, 900 Spring St., Silver Spring, Md. 20910
Savings & Loan League, U.S., 221 N. LaSalle St., Chicago, Ill. 60601
Savings Assn., National League of Insured, 1200 17th St., N.W., Washington, D.C. 20036
Scaffolding & Shoring Institute, 2130 Keith Bldg., Cleveland, Ohio 44115
Screen Mfrs. Assn., 110 N. Wacker Dr., Chicago, Ill. 60606
Sheet Metal & Air Conditioning Contractors, National Assn., 1611 N. Kent St., Arlington, Va. 22209
Sheet Metal Industry Promotion Plan, 909 The East Ohio Bldg., Cleveland, Ohio 44114
Shingle & Handsplit Shake Bureau, Red Cedar, 5510 White Bldg., Seattle, Wash. 98101
Shopping Centers, International Council of, 445 Park Ave., New York, N.Y. 10022
Siding Assn., Aluminum, 2217 Tribune Tower, Chicago, Ill. 60611
Slate Assn., National, 455 W. 23rd St., New York, N.Y. 10033
Society of American Registered Architects, 333 N. Michigan Ave., Chicago, Ill. 60601
Society of Construction Superintendents, 38 Park Row, New York, N.Y. 10038
Society of Real Estate Appraisers, 7 S. Dearborn St., Chicago, Ill. 60603
Society of the Plastics Industry, 250 Park Ave., New York, N.Y. 10017
Society of Wood Science & Technology, P.O. Box 5062, Madison, Wisc. 53705
Southern Cypress Mfrs. Assn., P.O. Box 16413, 1640 West Rd., Jacksonville, Fla. 32216
Southern Hardwood Lumber Mfrs. Assn., 805 Sterick Bldg., Memphis, Tenn. 38103
Southern Pine Assn., P.O. Box 52468, National Bank of Commerce Bldg., New Orleans, La. 70150
Southern Pine Inspection Bureau, P.O. Box 52468, New Orleans, La. 70150
Southern Pressure Treaters Assn., 3400 Northside Parkway, N.W., Sta. 7, Atlanta, Ga. 30327
Southern Woodwork Assn., 1909 Ardmore Rd., N.W., Atlanta, Ga. 30309

Specification Institute, American, 134 N. LaSalle St., Chicago, Ill. 60602

Specifications Institute, Construction, 1717 Massachusetts Ave., N.W., Washington, D.C. 20036

Spray Painting & Finishing Equipment Assn., National, P.O. Box 913, Toledo, Ohio 43601

Sprinkler & Fire Control Assn., National Automatic, 2 Holland Ave., White Plains, N.Y. 10603

Stained Glass Assn. of America, 3600 University Dr., Fairfax, Va. 22030

Standardization Society of the Valve & Fittings Industry Mfrs., 420 Lexington Ave., New York, N.Y. 10017

Standards Institute, American National, 1430 Broadway, New York, N.Y. 10018

Steel Construction, Inc., American Institute of, 101 Park Ave., New York, N.Y. 10017

Steel Deck Institute, 9836 W. Roosevelt Rd., Westchester, Ill. 60153

Steel Door Institute, 2130 Keith Bldg., Cleveland, Ohio 44115

Steel Joist Institute, 2001 Jefferson Davis Highway, Arlington, Va. 22202

Steel Structures Painting Council, 4400 Fifth Ave., Pittsburgh, Pa. 15213

Steel Window Institute, c/o Thomas Associates, Inc., Keith Bldg., Cleveland, Ohio 44115

Stone Assn., National Crushed, 1415 Elliott Pl., N.W., Washington, D.C. 20007

Stone Institute, Building, 420 Lexington Ave., New York, N.Y. 10017

Structural Clay Products Institute, 1750 Old Meadow Rd., McLean, Va. 22101

Structural Wood Fiber Products Assn., c/o MacLeary, Lynch, Gregg & Bernhard, 1625 K St., N.W., Washington, D.C. 20006

Stucco Mfrs. Assn., Inc., 15926 Kittridge St., Van Nuys, Calif. 91406

Surveyors Institute, Construction, 420 Lexington Ave., New York, N.Y. 10017

Swimming Pool Institute, National, 2000 K St., N.W., Washington, D.C. 20006

Tape Council, Pressure Sensitive, 1201 Waukegan Rd., Glenview, Ill. 60025

Technical Assn. of the Pulp & Paper Industry, 360 Lexington Ave., New York, N. Y. 10017

Terrazzo & Mosaic Assn., National, 716 Church St., Alexandria, Va. 22314

Testing Materials, American Society for, 1916 Race St., Philadelphia, Pa. 19103

Tile Contractors Assn. of America, 112 N. Alfred St., Alexandria, Va. 22209

Tile Council of America, 800 Second Ave., New York, N.Y. 10017

Tile Institute, Facing, 333 N. Michigan Ave., Chicago, Ill. 60601

Timber Construction, American Institute of, 1100 17th St., N.W., Washington, D.C. 20036

Title Assn., American Land, 1725 Eye St., N.W., Washington, D.C. 20006

Theatre Equipment & Supply Mfrs. Assn., 1270 6th Ave., Rockefeller Center, N.Y. 10020

Underground Engineering Contractors Assn., 8615 Florence Ave., Downey, Col. 90240

Underwriters, American Institute for Property & Liability, 270 Bryn Mawr Ave., Bryn Mawr, Pa. 19010

U.S. Copper Assn., 50 Broadway, New York, N.Y. 10004

U.S. National Committee for C I B (Construction), 2101 Constitution Ave., Washington, D.C. 20418

U.S. Savings & Loan League, 221 N. LaSalle St., Chicago, Ill. 60601

Upholstery Fabric Mfrs. Assn., 122 East 42nd St., New York, N.Y. 10017

Urban Land Institute, 1200 18th St., N.W., Washington, D.C. 20036

Valve & Fittings Industry Mfrs., Standardization Society of the, 420 Lexington Ave., New York, N.Y. 10017

Ventilating Institute, Home, 1108 Standard Bldg., Cleveland, Ohio 44113

Vermiculite Assn., 527 Madison Ave., New York, N.Y. 10022

Vermiculite Institute, 208 S. LaSalle St., Chicago, Ill. 60604

Vinyl Fabrics Institute, 60 East 42nd St., New York, N.Y. 10017

Vinyl & Rubber Flooring Division, Rubber Mfrs. Assn., 444 Madison Ave., New York, N.Y. 10022

Wallcoverings Council, 969 Third Ave., New York, N.Y. 10022

Wall Paper Institute, 969 Third Ave., New York, N.Y. 10022

Water Conditioning Assn., International, 323 West Wesley St., Wheaton, Ill. 60187

Water Conditioning Foundation, 1780 Maple St., Northfield, Ill. 60093

Water Conditioning Research Council, 325 West Wesley St., Wheaton, Ill. 60187

Waterproofers Assn. Inc., Building, 60 East 42nd St., New York, N.Y. 10017

Water Well, National, 88 E. Broad St., Columbus, Ohio 43215

Water Works Assn., American, 2 Park Ave., New York, N.Y. 10016

West Coast Lumber Inspection Bureau, 1750 S.W. Skyline Blvd., P.O. Box 25406, Portland, Ore. 97225

Western Forest Industries Assn., 1500 Southwest Taylor, Portland, Ore. 97205

Western Red Cedar Lumber Assn., 707 Joseph Vance Bldg., Seattle, Wash. 98101

Western Red & Northern White Cedar Assn., P.O. Box 2786, New Brighton, Minn. 55112

Western Wood Products Assn., 700 Yeon Bldg., Portland, Ore. 97204

Window Institute, Steel, c/o Thomas Associates, Inc., Keith Bldg., Cleveland, Ohio 44115

Wire Reinforcement Institute, 5034 Wisconsin Ave., N.W., Washington, D.C. 20016

Wire Weavers Assn. (screen), 441 Lexington Ave., Suite 608, New York, N.Y. 10017

Women in Construction, National Assn., (NAWIC), 1000 Vermont Ave., N.W., Washington, D.C. 20005

Wood Flooring Institute of America, 210 N. Wells St., Chicago, Ill. 60606

Wood Preservers Assn., American, 1012 Fourteenth St., N.W., Washington, D.C. 20005

Wood Preservers Institute, American, 2600 Virginia Ave., N.W., Washington, D.C. 20017

Wood Science & Technology, Society of, P.O. Box 5062, Madison, Wisc. 53705

Woodwork Institute, Architectural, Chesterfield House, Suite A, 5055 S. Chesterfield Rd., Arlington, Va. 22206.
Woodwork Mfrs. Assn. Inc., National, 400 W. Madison St., Chicago, Ill. 60606
Woodworking Machinery Distributors Assn., 1900 Arch St., Philadelphia, Pa. 19103
Woodworking Machinery Mfrs. of America, 1900 Arch St., Philadelphia, Pa. 19103

Zinc Institute, Inc., 292 Madison Ave., New York, N.Y. 10017

Index

A

Abstract of Bids Opened, form, 59
Abstract of Bids or Informal Proposals, form, 60
Accident prone, new personnel more, 29
Accident: (also see OSHA)
   causes, 163
   reports, 164
Acoustical ceiling tile:
   irregularities in installation, 184–85
   proper installation, 185
Adobe brick (see Brick, adobe)
AGC Operating Engineers Training School, 143
Agreements, joint venture, 65, 69, 70–72
Algae stains, removal, 179
Aluminum stains, removal, 186–87
Aluminum surfaces, primer, 187
Annual Summary Sheet (equipment) forms, 130
Applicant:
   selection and evaluation, 41
   telephone check form, 42
Application for Employment form, 36
Application for Office Position form, 39–40
Application for Position form, 37–38
Aptitude, determining mechanical, 41
Architectural Research services, 191
Asphalt tile, softening problem, 183
Associations (see Appendix), 199–215

B

Birds roosting on buildings, prevention, 186
Blasting, OSHA requirements, 161
Boulders, best method of removal, 145
Brick, adobe, manufacture, 183–84
Brick, stains, removal, 179–80
Building interiors, proper environmental conditions, 122
Building trades specialization, 53
Burning bars for concrete cutting, 182
Bushhammer tools, 197
Bushhammered concrete finish:
   how to attain, 178
   proper technique, 178–79
Business forms: (also see Forms)
   essentials needed for layout, 87–88
   you can design your own, 87
Business information sources, 192

C

Cabinetwork, purchasing, 121
Carriers, what can be done to avoid delivery delays, 58
Ceiling tile, acoustical, installation, 184–85
Change orders, use, 96
Chart:
   organization, 30
   policy making, 30
Charts, graphs and forms, 195
Check list and score sheet for road test in traffic, 141–42
Civil engineering, use of laser, 144
Claims:
   accident forms, 165–68
   damage or injury, 169–70
Color coding, benefits, 86
Communication:
   avoiding costly errors, 30
   with personnel, 29
Computer:
   application to designing, 77
   application to job scheduling, 76
   CPM by use, 76
   drafting, 77
   engineering applications, 78
   estimating application, 75
   help for subcontractor, 76
   retrieval of data, 78
   role in management, 78

Computer (*cont.*)
   time-sharing, 75
Concrete:
   best method of breaking, drilling, 145
   bushhammered finish, 178
   clay stains, removal, 179
   curing procedures, 177–78
   cutting method, 181–82
   exposed aggregate surface, 187–88
   finishing, 176–77
   five methods of protecting while curing, 177
   improved forming, 178
   problems, 196
   proper forms save grinding, 178
Concrete floors, prevention of dusting, 176–77
Concrete forms and form liners, 196
Confirmation of telephone quotation forms, 62–63
Contract, joint venture agreement, 65, 69, 70–72
Contractor, what size firm can best utilize CPM, 52
Contractors, loose-leaf desk book, 191
Control systems, use of visual, 79
Construction:
   cost estimate, 68
   daily report forms, 99–100
   problem solving methods, 173
   report forms (Lefax), 112–13
   stress and pressure in jobs, 28
Construction site (see Job site)
Cost distribution (see Job cost distribution)
Cost:
   planning for lower construction, 47
   record form (equipment), 130
CPM:
   advantages of use, 93
   computer not always necessary, 51
   decision-making process, 52
   displays, charts and boards, 80
   how to apply, 50, 51
   school for training, 51
   type of projects applicable, 52
   who should be using, 51–52
Cracks and open joints in interior woodwork, 176
Cracks in terrazzo, 180
Cracks in terrazzo floors, prevention, 180
Critical Path Method (see CPM)
Curing agents to avoid concrete problems, 183

D

Daily construction report and progress report forms, 110
Daily construction report forms, 100, 114
Daily job report, 101
Daily material report, 100
Daily report form, 115
Damage and injury forms, 169
Delays caused by carriers, how to avoid, 58
Deliveries (see Carriers)
Demolition, 161
Displays, how to make, 80
Drafting aids, pressure-sensitive transfers, 79
Drafting by computer, 77
Drawings, shop approval, 64
Driver tests, 137–42

E

Education (see Schools, equipment operation)
Electro-osmosis dewatering, 196
Employee: (also see Personnel)
   attitude toward company, 133
   multipurpose time and job forms, 106
   safety training, 162
   time and job forms, 107
   training on equipment use, 131, 133
   training on office machines, 131
Employment application forms, 36
Encyclopedia of American Associations, 189
Equipment: (also see Tools and specific kinds)
   boulder splitter, 182
   charge system, 131
   charges to project form, 132
   handling methods, 173–74
   hazards of unsafe equipment, 163
   modernization, 134
   prolonging life of construction, 127
   reducing losses, 134–35
   theft prevention program, 135
   torch for cutting through concrete, 181
Equipment, construction:
   daily cost record form, 129
   depreciation schedule, 128
   drilling or piercing, 144
   establishing maintenance schedule, 128
   helicopters, 143–44
   importance of periodic inspections, 128
   improving efficiency, 134
   job cost records, 131

INDEX

Equipment, construction (*cont.*)
    laser beam instrument, 144
    list of essential records, 131
    maintenance, 127–28
    new requirements for OSHA compliance, 155
    operator schools, 143
    proper check-out, 133
    proper use and operation, 128
    psychological factors affecting life, 133
    replacement program, 128
    scheduling system, 128
    special records file, 127
    splitter for breaking boulders, 145
    use of claw, 144
    use of radial saw (also see Radial saw), 145
Equipment life, psychological factors affecting, 133
Equipment, new, proper training, 131
Equipment operator school, 143
Equipment program of preventive maintenance, 128
Equipment, trucking:
    driver training program, 136
    maintenance procedure, 143
    purchasing, 136
    reducing cost, 136
    replacement program, 136
    testing of operators, 136
    timing of purchases, 136
Excavation shoring, OSHA requirements, 159
Expediter, importance of, 58
Extra work order forms, 98
Extra work orders, use, 96

F

Fasteners, 196
Fire-retardant materials (see Wood, Tile, acoustical)
Floors:
    prevention of cracks in terrazzo, 180
    stains on terrazzo, 179
Follow-up summary forms, 49
Forms:
    abstract of bids opened, 59
    abstract of bids or informal proposals, 60
    annual summary sheet (equipment), 130
    applicant telephone check, 42
    application for employment, 36
    application for office position, 39–40
    application for position, 37–38

Forms (*cont.*)
    avoid verbal orders (A.V.O.), 90
    check list and score sheet for road test in traffic, 141–42
    claim for damage or injury, 169–70
    confirmation of telephone quotation, 62, 63
    construction cost estimate, 68
    construction report (Lefax), 112–13
    daily construction report, 100
    daily construction report (Lefax), 114
    daily construction report and progress report, 110
    daily job report, 101
    daily material report, 100
    daily report, 115
    developing your own business, 23
    employee time and job, 107
    employee time and job multipurpose, 106
    equipment charges to project, 132
    equipment cost daily record, 129
    extra work order, 98
    follow-up summary, 49
    for internal control, office, 23
    installment note, 111
    investigation report of motor vehicle accident, 165–66
    job record sheet, labor, 103–04
    labor report, 102
    memorandum of telephone calls, 90
    monthly time and cost record (equipment), 130
    operator's report of motor vehicle accident, 167–68
    position analysis, 34–35
    proposal and acceptance, 108
    purchase and hire progress report, 66–67
    purchase order, 109
    record of bidders, 61
    report of accident other than motor vehicles, 171
    room finish schedule, 94
    selection and evaluation summary, 44
    sender's application for recall of mail, 89
    statement of supervisory expectancies, 32–33
    standardized test, traffic and driving knowledge, 137–40
    subcontractor agreement, 54–55
    telephone check on executive applicant, 43
    time and job ticket (3-part form), 105
    waiver of lien, 111
    work order—change order, 97

Forms, OSHA:
   log of occupational injuries, illnesses (OSHA 100), 150
   summary occupational injuries and illnesses, (OSHA 102), 151
   supplementary record of occupational injuries and illnesses (OSHA 101), 152

G

Glass, how to determine tempered, 175
Glass surfaces, how to clean discolored exterior, 184
Ground water:
   control, 182
   electro-osmosis method, 183
   removal, 182
   wellpoint dewatering method, 182–83
Guarding projects, 134

H

Helicopters, use in construction, 143
Helicopters, use in pick-up and delivery, 144
Hiring interview procedure, 31

I

Illness on job (see OSHA)
Illustrations (see Forms)
Injuries on job (see OSHA)
Interior woodwork, cracks and open joints, 176
Ironwork (see Steel)

J

Job cost distribution, 85
Job cost distribution stamp, 86
Job description form, 35
Job description, written, 31
Job progress reports, 65
Job problems:
   check list, 95
   close supervision reveals, 95
   corrective action, 95
   how to prevent accumulation, 95
   inspection of deliveries, 95
   inspection of vendor manufactured items, 95

Job problems (*cont.*)
   shop drawings approval, 95
   subcontractor's, 95
Job record sheet, labor form, 103–04
Job report form, daily, 101
Job-site:
   avoiding errors, 91
   clean air requirements (also see OSHA), 155–56
   corrective action for delays, 92
   electrical requirements of OSHA, 157
   hazards to children, 136
   noise level requirement (also see OSHA), 155
   off-hours protection, 135
   safety consciousness, 162
   use of guard dogs, 135
   use of helicopters, 143
Job superintendents, responsibility, 47
Joint venture agreements (see Agreements, joint venture)

L

Labor report form, 102
Labor reports, 99
Laborers (see work force), 26
Laser beam:
   advantages over transit, 144
   uses in construction, 144
Lumber:
   high cost of below-grade, 120–21
   how to avoid shrinkage cracks, 122
   inspection, 120, 122
   moisture absorption on site, 122
   placement of materials at job site, 122
   purchasing, 120

M

Machines, office:
   proper use and operation, 128
   replacement program, 128
   setting up depreciation schedule, 128
Magnetic symbols for display boards, 80
Mail, sender's application for recall form, 89
Management:
   attitudes, 93
   consultants, need for, 48
   development programs, 30

INDEX
221

Management personnel:
  checking references, 28
  creativity and innovation, 28
  determining qualifications, 27
  experience, 27
  how to determine mechanical skills, 27
  how to find, 26
Managers, qualifications, 26
Masonry:
  materials to control moisture below grade, 175-76
  stains, removal, 179-80
Masonry surfaces:
  removal of tobacco stains, 187
  water seepage control, 175-76
Masonry walls and floors, how to control water seepage, 175-76
Mastic, avoid use of "cut-back," 183
Material report form, daily, 100
Materials:
  allowance for shrinkage, 124
  checklist for receiving on job, 123
  expediting, 117
  fallacy of cheating on quality, 123-24
  placement on job site, 122
  preventing loss, 117, 119
  protecting from theft, 119, 135
  protecting from weather damage, 119
  protection on job, 118
  purchasing, 117
  reducing loss, 134-35
  safeguard in purchases, 118
  samples to be submitted, 118
  special handling of cement, lime and plaster, 119
  storage on job, 118
  unauthorized substitution, 123
Materials waste:
  corrective actions, 120
  survey best approach, 120
Materials and supplies:
  acceptance, 91
  problems, 91
Methods engineering for lower operating cost, 80, 195
Methods engineering studies, equipment, 81
Millwork:
  high cost of below-grade, 120-21
  inspection, 120
  purchasing, 120
Moisture meter, use to determine moisture content of lumber, 122

Mortar stains on aluminum, removal, 186-87
Motivation:
  checklist, 27
  human, 26
  vs. manipulation, 26-27
Motor vehicle accident forms, 165-68
Motors, electrical, power requirements, 146

N

Note, installment, form, 111

O

Occupational injuries and illness forms, 150-52
Occupational Safety and Health Act (see OSHA)
Office positions, application, 31
Orders, avoid verbal, form, 90
Organization:
  analyzing objectives, 21
  charts, 30
  degree of specialization, 22
  delegation of work load, 22
  forms and records, 22
  personnel, 25
  procedure manual, 23
OSHA:
  accident reports, 153
  air and oxygen requirements for tunnels, 160
  appeal procedures, 148
  assistance through SBA, 153
  barricade requirement, 156
  blasting regulations, 161
  citations for violations, 148
  clean air requirements, 155-56
  communications in tunnels, 160
  compressed air use, 157
  defective equipment, 156
  demolition precautions, 161
  effective dates, 162
  effects of noise on employees, 155
  electrical requirements on job site, 157
  electrical switches on power equipment, 157
  employee complaints, 148
  enforcement of law, 148
  excavation, safety requirements, 159
  explosives (see blasting), 161
  fender requirements for rubber-tired vehicles, 159
  fire protection requirement, 156

OSHA (*cont.*)
    flooring requirements for high rise, 160
    horn requirements for bidirectional vehicles, 159
    "headache ball" restrictions, 161
    industry standards, 148
    inspection of machinery and equipment, 158
    inspection of workmen's tools, 157
    ladder specifications, 158
    length of time to report fatality, 153
    load limits on floors, 156
    log of cases, 149–50
    materials handling regulations, 156
    materials storage, 156–57
    motor vehicle safety rules, 158
    monetary grants for states, 154
    penalties for violations, 148
    posting of notices required, 148, 153
    publications available, 153–54
    recording procedures for injuries and illnesses, 149
    records on inspection of machinery and equipment, 158
    records required, 149
    records retention, 149
    reports by employees, 149
    reports of violations, 148
    responsibility of contractor, 155
    responsibility of subcontractor, 155
    revisions, 147
    rules for parking equipment, 158–59
    safe use of compressed air, 160
    safety belts, 159
    safety inspector's visits, 148
    safety net requirements, 160
    scaffolds specifications, 158
    signs and signals required, 156
    sources of information, 148
    state vs. federal enforcement, 154
    steel erection safety requirements, 159–60
    trench shoring requirements, 159
    underground openings safety, 160
    ventilation doors, 160
    waste disposal, 157
    who is affected, 147
    who may examine logs, 149

P

Painting problems, solutions, 185
Particle board, use for shelving, 122
Patrolman for protection of job site, 135
Personnel:
    avoidance of humiliation, 29
    basic needs, 29
    communication with all, 29
    fitting new, into organization, 29
    how to keep, 29
    new are more accident prone, 29
    organization, 25
    problem of hiring skilled, 25
    technique of handling, 25
Pipe laying without excavating, 144
Planning:
    conference technique, 50
    critical path method, 50
    techniques, 48
Plat map, as an aid to materials placement, 123
Plate glass, cleaning exterior surface, 184
Pocket size data systems:
    Lefax's, 106
    Problem Solution Associates, 106
    Frank R. Walker's, 106
Policy chart making, 30
Position analysis form, 34–35
Problem Solution Associates, 190, 191
Problem solving in construction, 173, 190
Problem-solving techniques, 173–74
Problems (see Job problems)
Procedure manual:
    setting up for organization, 30
    value, 23
Proposal and acceptance form, 106, 108
Project:
    internal planning and control, 47
    job-site control, 91, 106
    reports, 65, 106
    scheduling (see CPM scheduling)
    type applicable to CPM, 52
Pocket system of forms, 56, 106
Protective coatings (see Lumber and Steel)
Purchase order forms, 109, 106

R

Radial saw:
    alignment procedures, 145
    operator instructions, 145–46
    proper maintenance, 146
    safe operation, 146
Record keeping:
    accidents, 96

INDEX 223

Record keeping (*cont.*)
   use of camera, 96, 98
   use of daily diary, 96, 106
   rules, 96
Responsibility, determining individual, 28
Responsibility vs. authority, 30
Reference sources:
   Basic Reference Sources, 192
   building code organizations, 190
   computer application to construction, 194
   critical path and other scheduling methods, 195
   Guide to American Directories, 192
   How and Where to Find the Facts, 192
   How and Where to Look It Up, 192
   Kelley's Directory of Merchants, Manufacturers and Shippers, 192
   management development, 197
   methods engineering services and equipment, 195
   Norman Foster books, 191
   OSHA, 197
   Prentice-Hall, Inc., 193–94
   Problem Solution Associates, 106, 190, 191
   Solution to Problems in Building and Construction, 190, 191
   safety literature, 197–98
   Sources of Business Information, 192
   Standard Rate and Data Guide, 192
   stormproofing, 194
   Superintendent of Documents, 192
   Sweet's Catalog Service, 191
   Thomas Register of American Manufacturers, 191
   university research bureaus, 190
   visual aids sources, 194–95
   Walker, Frank R. & Co., 191
References:
   checking applicant's, 28
   how to evaluate applicant's, 28
Rock removal simplified, 144
Room finish schedule form, 94
Roosting of birds on building, preventing, 186

S

Safety: (see OSHA)
   causes of accidents, 163
   psychological factors, 164
   record keeping, 96
SBA, assistance in OSHA compliance, 153

Schedules:
   adherence to by subcontractors, 58
   adherence to by suppliers, 58
   maintaining job, 57, 92
Scheduling (see CPM)
Schools:
   craft, 143
   heavy equipment operation, 143
Shipments (see Carriers)
Small Business Administration (see SBA)
Soil stabilization, 182–83
Sources of construction information, 189–98
Steel:
   preparation of surface, 125
   protective coatings, 124–25
Stone working tools, 197
Subcontractor:
   addenda to clarify questionable phases of job, 56
   adequate written contract, 53
   advantages, 53
   agreements, 57
   communications with, 57
   importance of doing work thoroughly, 58
   how to obtain cooperation, 53
   keeping informed of schedule, 93
   method of payment, 57
   supervision of labor, 53
Supervisors' training in conservation of materials, 120
Supervisory expectancy, statement, form, 32–33
Suppliers, keeping informed of schedule, 93
Stains:
   removal from aluminum, 186–87
   removal from brick, 179–80
   removal from concrete, 179, 187
   removal from glass, 184
   urine on carpet, treatment, 185
Systems, magnetic visual control, 80

T

Telephone calls, memorandum form, 90
Telephone check of applicants, form, 41
Tempered glass verification, 175
Tensions:
   how to cope with them, 41
   ways to avoid, 46
Terrazzo:
   floor construction, 181
   prevention of cracks, 180

Terrazzo (*cont.*)
   repair of cracks, 180
   removal of stains from floors, 179
Tests:
   comprehension, 41
   dexterity, 41
   mechanical aptitude, 41
Theft of tools, equipment and materials, 135
Tile, acoustical, fire-retardant, 118
Tile, asphalt, softening, 183
Time and cost record form, 130
Time and job ticket form, 105
Time-lapse photography:
   equipment, 81–82, 195
   reducing construction costs, 82
   use in construction, 81
   use in safety engineering, 82

T

Tobacco stains, removal from masonry surfaces, 187
Tools: (also see Equipment, OSHA, Radial saws)
   construction, 127
   preventing theft, 135
   reducing loss, 134–35
   unusual tools and equipment, 197
Trucking equipment (see Equipment, trucking)

U

Underground piercing equipment, 144
Urine stains on carpets, treatment, 185

V

Vehicle parking, wheel chocks required by OSHA, 158
Video tape recorder and monitors, use in construction, 83, 195
Visual aids, 79, 194, 195

W

Waiver of lien form, 111
Watchman, protection on job site, 135
Water (see Ground water)
Water seepage problems, 175
Water seepage through masonry below grade, how to control, 175–76
Wiseman Personnel Classification Test, 27
Wood:
   chemical treatment, 117–18
   fire-retardant, 117
   research on fireproofing, 118
   use in construction, 117
Wood products: (also see Lumber and Millwork)
   correct priming, 122
Woodwork:
   compression set problem, 176
   importance of proper moisture content, 122
Woodwork, moisture content, see map, 121
Woodworking operations, improving quality and efficiency, 145
Work force:
   check list, 93
   provide full information, 93
Work order forms, 97, 106
Work orders, use in construction, 96